Technician Manufacturing Technology IV

Technician Manufacturing Technology IV

M. Haslehurst D.I.C.

Formerly Dean of the Faculty of Technology and Science
North Cheshire College

HODDER AND STOUGHTON
LONDON SYDNEY AUCKLAND TORONTO

British Library Cataloguing in Publication Data
Haslehurst, Maurice
  Technician manufacturing technology, IV.
  1. Machine-shop practice
  I. Title
  621.7'5   TJ1125

  ISBN 0–340–22098–8

First published 1979
Reprinted 1982

Printed and bound in Great Britain for Hodder and
Stoughton Educational, a division of Hodder and
Stoughton Limited, Mill Road, Dunton Green, Sevenoaks, Kent,
by Richard Clay (The Chaucer Press), Ltd., Bungay, Suffolk

Preface

As a result of the establishment in March 1973 of the Technician Education Council (TEC) the National and Higher National Certificate, and the CGLI technician engineering courses have been replaced by TEC Certificate and Higher Certificate courses. This process of rationalisation is now well advanced, although the position in the area of National and Higher National Diploma courses is perhaps not yet so clear. The introduction of the TEC courses has created a need for a new range of text books to cover the TEC standard units (syllabuses) of study.

This book is designed to cover the learning objectives of the standard unit 77/423 of Manufacturing Technology at Level IV in the A5 Programme for Mechanical and Production engineering. Level IV is the first of the two levels of study for the A5 TEC Higher Certificate (THC), although it should be appreciated that some lower level units of study can be credited for the THC award.

Each chapter specifically covers the learning objectives set out in TEC unit 77/423, the objectives being listed at the end of each chapter. Some extra material is added where the author feels that a greater depth of treatment than that indicated might be desirable. All chapters, with the exception of Chapter 1, are illustrated by the use of line diagrams. These are intended to be sufficiently simplified so that the principle under discussion is more readily understood, and that may be relatively easy for the student to reproduce. At the time of writing, no TEC standard unit has been published for Level V of the A5 THC programme for Mechanical and Production engineering. However, the author hopes that his previously published text book 'Manufacturing Technology', which has been used by HNC students for some years, may continue to be useful to students studying this subject at Level V.

Finally, I would like to thank everyone who has assisted me in any way with the preparation of this book, particularly Mr. K. R. C.

Moore of J. P. Udal Ltd., Birmingham, and Mr. P. Haden of The Silvaflame Co. Ltd., Walsall, whose companies have kindly allowed me to use photographs which illustrate Chapter 1. My deepest thanks go to my wife, Gwyneth, for typing the manuscript and continuing to encourage me.

M.H.
Chelford

Contents

Chapter 1
Safe Working Practice

The Health and Safety at Work (etc.) Act 1974 came into existence on 1st April 1975. The Act provides for:
1. One comprehensive system of law dealing with the health and safety of virtually all people at work.
2. Protection for the public where they may be affected by the activities of people at work.
3. The setting up of a Health and Safety Commission and an Executive responsible to Ministers. These bodies will administer the legislation on health and safety matters.

This Act is an enabling one. That means that it enables the Commission, with the approval of the appropriate Minister, to make changes to the existing law without the need to gain Parliamentary approval. Hence changes can quickly be brought about. Most of the current health and safety legislation, such as the Factories Act 1961, will remain in existence until such time as it is replaced by: (*a*) new and improved regulations made under the new Act, and (*b*) approved 'Codes of Practice'.

1.1 Safety regulations

The Factories Act 1961 requires all dangerous parts of machinery to be securely fenced. Over many years interested parties, such as machine and guard manufacturers, users, the factory Inspectorate and the British Standards Institution have made great efforts to develop effective methods of complying with this particular piece of legislation. For example, BS 5304 has been drawn up and offers a code of practice for safeguarding of machinery.

(*a*) **Press working.** Specific requirements for mechanical power presses are contained in the Power Presses Regulations 1965 and the Power Presses (Amendment) Regulations 1972 which deal in particular with the training of press-tool setters, the daily inspection of press-guards, and the periodical thorough examination of guards and machines. The tools of power presses are known to be amongst the most dangerous parts of machinery in common use in a factory, and statistics show that tool-setters as well as operators sustain accidents. The above Regulations state that persons who prepare power presses for use must be 18 years or over, and receive training which includes suitable and sufficient practical instruction on:

  (i) Power press mechanisms.
 (ii) Safety devices.
(iii) Accident causation and prevention.
(iv) The work of the tool-setter.
 (v) Tool design.

TEC Unit U75/035 'Manufacturing Technology II' covers guard systems for power presses, but it will be worth while to look again at this important aspect of press safe working practice. However, it must be emphasised that irrespective of guarding, it is essential to maintain safe working conditions and safe operating methods. These are the joint responsibilities of employer and employee.

Guarding methods. Press Tools may be made safe in themselves, such that there is insufficient space anywhere on the tool-set for the entry and trapping of fingers. Such tools are known as 'closed tools' and are mostly used for blanking operations. The three principal ways in which guards are applied as safety devices are as follows:

1. *Fixed guards*, which at all times prevent access of any part of the body to the tool area. This type of guard should preferably always be used except when access of the operator's hands to the tool is essential. Figure 1.1 shows a typical fixed guard.

2. *Interlocked guards*, which have a movable gate fitted to an enclosing fixed guard. The gate is interlocked with the press mechanism in such a way that the press cannot be operated unless the gate is closed. The gate cannot then be opened by the operator until the press ram has returned to its usual stopping position and the tool is in the open position. Safe access can then be gained to the tool. This type of guard can be used on presses of all types and size. Figure 1.2 shows a typical interlocked guard.

3. *Automatic guards*, which push or pull the operator clear before trapping in the press tool can occur. These guards have limitations as to their use, such as the crankshaft speed exceeding certain limits, or the press stroke being inadequate. The automatic guard is designed to push or pull an operator's body, arm or hand clear of danger while the tool is descending. There are two types of automatic guard in general

Fig 1.1 Fixed guard

Fig 1.2 Interlocked guard

Fig 1.3 Automatic guard

use, viz., the push-away or rising-screen guard, and the pull-out guard. Figure 1.3 shows an automatic guard of the push-away type.

(*b*) **Abrasive wheels.** Specific requirements for machines using abrasive wheels are contained in the Abrasive Wheels Regulations 1970, supplemented by the Protection of Eyes Regulations 1974. The latter state that employers must provide eye protection for every employee who works in any of the specified processes, including grinding, and that a person provided with eye protectors or a shield must make full and proper use of it.

The Abrasive Wheels Regulations set out in detail certain precautions which must be taken in connection with the examination, storage, handling, selecting, mounting, truing and dressing of wheels; training of persons to mount wheels, guards and work-rests where provided. Speeds of wheels and spindles is covered, over-speeding being the most common cause of the bursting of abrasive wheels. The Regulations require that every abrasive wheel having a diameter of more than 55 mm shall be marked with the maximum permissible speed in revolutions per minute, and that the working speed of every grinding machine shall be specified in a notice attached to the machine.

Guarding methods. Guards must be provided and kept in position at every abrasive wheel unless exceptionally the nature of the work precludes its use. The functions of a wheel guard are to contain the wheel if it bursts, and to prevent the operator touching the wheel. The development of high-speed grinding in recent years has increased the technical difficulties of satisfying the first function. Generally then, a wheel should be enclosed to the greatest possible extent, the guard opening being as small as practicably possible. As the wheel wears, the guard should be capable of adjustment in order to maintain maximum protection. A variety of grinding machines are in existence from portable to large precision machines, having single or several wheels. Figure 1.4 shows a guard suitable for one of this large range of

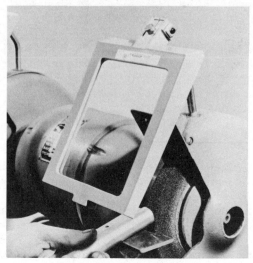

Fig 1.4 Grinding-wheel guard

machines namely an off-hand tool grinder. A wheel-rest is also incorporated.

(c) **Milling machines.** In addition to the Factories Acts, the Horizontal Milling Machines Regulations 1928 should be consulted in order to appreciate the hazards associated with the milling process, and the precautions to be taken. The process makes use of revolving toothed cutters which are particularly dangerous. Analysis of accidents show that the majority of injuries are sustained on horizontal machines when repetition work is being undertaken, and that over one third of accidents occur when the workpiece is being loaded or unloaded. This suggests that improvements are required in the efficiency of operation of milling machine guards.

Clearly, in addition to efficient guards, safe working practice and safe working conditions are essential, and operators must be trained in these aspects. Sloppy clothing and long hair is dangerous, and where necessary eyes should be protected against flying swarf.

Guarding methods. Every milling cutter must be provided with an efficient guard. The Regulations call for complete enclosure of the cutting area, and this should be achieved even at the expense of ease of access to the cutter. Guards are often designed to be adjustable in order to accommodate varying sized cutters, and the modern tendency is to introduce transparent panels which permit observation of the milling operation. Ideally, an interlocking guard should be used which ensures that the cutter can be in motion only when the guard is closed. Figures 1.5 and 1.6 illustrate modern guards for horizontal and vertical milling operations respectively.

1.2 Health and Safety Act 1974

The introduction to this chapter very briefly outlines the main provisions of this Act. However in this section it will be necessary to examine it in more detail with particular respect to safe working practice of manufacturing processes.

The Act is in three parts, but only Part I contains the *Health and Safety at Work* provisions. The Parts are sub-divided into sections, and there are 54 of these in Part I. These are again divided into numbered sub-sections and paragraphs. Hence, reference 23(5)(*b*) refers to paragraph (*b*) of sub-section (5) of section 23 of the Act. It would be impractical to enumerate the contents of each here, as a copy of the Act runs to 117 pages, but we will identify those parts relevant to safe working practice.

SECTION 2 places a general duty on employers to ensure the health, safety and welfare at work of their employees. For example,

Fig 1.6 Vertical milling machine guard

ref: 2(2)(*a*) 'the provision and maintenance of plant and systems of work that are, so far as is reasonably practicable, safe and without risks to health'.

SECTION 3 places a general duty on employers to ensure that their activities do not endanger persons not in their employment.

SECTION 4 places a general duty on persons in relation to those who are not their employees, to ensure that plant made available to them is safe and without risk to health.

SECTION 6 places duties on anyone who designs, manufactures, imports, supplies or installs an article or substance for use at work, to ensure that it is safe and without risk to health.

SECTION 7 places duties on employees to take reasonable care to ensure that they do not endanger themselves, or anyone else, who may be affected by their work activities. For example, ref: 7(*a*)——duty of every employee——'to take reasonable care for the health and safety of himself and of other persons who may be affected by his acts or omissions at work'.

SECTION 8 places a duty on all persons not to misuse anything provided in the interests of health or safety purposes under a statutory requirement.

The implications of this are clear. The Act places general duties and obligations on everybody concerned with manufacturing processes, including employers, employees, suppliers, etc., to ensure the safety and health of all. The penalties for evading these duties are strict, and include imprisonment.

Learning Objectives 1
The learning objectives of this section of the TEC Unit are:

1. Understands and is aware of safe working practice.

1.1 Explains the relevance of safety regulations applying to (*a*) press working (*b*) abrasive wheels (*c*) milling machines.
1.2 Explains the implications of the appropriate sections of the Health and Safety Act 1974 to the manufacturing processes in the unit.

Exercises 1

1   Refer to copies of the following documents, and particularly read the sections relating to your own industry:
   (*a*) Factories Act 1961.
   (*b*) Health and Safety at Work Act 1974.
   (*c*) BS 5304: Safeguarding of Machinery.
2   (*a*) What are the three principal ways in which guards are applied as safety devices to power presses?
   (*b*) Sketch one type of power press guard.
3   (*a*) What are the two main functions of an abrasive wheel guard?
   (*b*) Sketch a suitable guard for an off-hand tool grinder.

4 (*a*) What is the main purpose of a milling machine guard?
    (*b*) Sketch a suitable guard for either a horizontal milling machine, or a vertical milling machine.
5 (*a*) What are the main provisions of the Health and Safety at Work Act?
    (*b*) Summarise the contents of the main parts of the Act which refer to safe working practice.
6 Refer to copies of the following documents and read the main provisions:
    (*a*) Power Presses Regulations 1965.
    (*b*) Power Presses (Amendment) Regulations 1972.
    (*c*) Abrasive Wheels Regulations 1970.
    (*d*) Protection of Eyes Regulations 1974.
    (*e*) Horizontal Milling Machines Regulations 1928.

Chapter 2
Metal Forming

Presses are commonly used in industry for the cold working of metal objects into a variety of shapes. Most of this type of work is carried out upon ductile metal in sheet or strip form of relatively thin section. It represents an important part of manufacturing industry, being used for the cheap production of large quantities of components, such as motor car bodies, electric motor parts, domestic electrical appliances parts, etc.

Orthodox presses (usually vertical) are used, and may be hydraulic or pneumatic powered, or may be mechanical crank presses. For very light work, hand presses such as fly presses, are satisfactory. The metal, if in coiled strip form, may be fed automatically into the press tool by power rolls or power slide, or may be hand fed by the operator. If the metal is in some other form, such as a sheet or partially formed shape, it may be located in the tool by mechanical hands which have gripping fingers, or locating pans which drop the metal part into the correct position. Again, an operator(s) may hand feed the part. The mechanical feed or location devices must of course be synchronised to operate every time the press ram lifts the top tool clear of the bottom tool. Where the press is hand fed, stringent safety precautions must be taken to ensure that the operators' hands cannot be trapped in the press tool. Efficient guards must be provided which are completely foolproof, and it should always be remembered that a press is potentially a very dangerous machine. (See Section 1.1.)

The three basic techniques of cold working sheet metal in press tools mounted on presses are *blanking*, *bending* and *drawing*. The first two techniques are covered in TEC Unit U75/035 'Manufacturing Technology II' and this chapter will therefore be confined to the drawing process of forming a hollow object under a press from metal initially in the form of a relatively thin blank.

2.1 Drawing tool
Figure 2.1 shows a simple *drawing tool* in which a blank is being drawn into a cylindrical cup. The work is shown in a partially drawn state, the punch not yet having 'bottomed'.

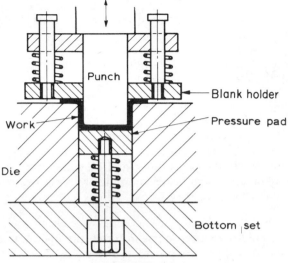

Fig 2.1 Drawing tool

The important elements of a drawing tool are shown labelled in Figure 2.1, and are as follows:

Punch. Made from hardened and ground alloy steel, and located exactly on the die axis. The circular punch descends under the action of the press-ram and carries the metal blank into the die aperture.

Die. Made from the same hardened metal as the punch. The die receives the blank as it is forced into position by the punch. A suitable lubricant must be used to assist the process.

Blank holder (See Section 2.5). Figure 2.1 shows the blank holder springs compressed. It is designed to press down on the blank just before the punch starts to impinge on the top face of the blank. The blank holder presses the metal against the die face so that the blank is *ironed out* as it is drawn over the radiused edge of the die throat.

A more suitable, but expensive way of applying this pressure is to use a double-action press.

Pressure Pad (See Section 2.7). The function of the *pressure pad* (or die cushion) is to support the work as it descends into the die under the action of the punch, and also to act as a component ejector. The pressure on the bottom of the cup, maintained by the cushion, keeps it flat and prevents puckering.

If insufficient pressure can be created by the simple device shown in

Figure 2.1, then this can be increased in practice by a heavier arrangement mounted under the press, the pressure being supplemented by large rubber blocks.

Sets. The die is fastened to the bottom set, which in turn is bolted or clamped to the press bed. Sometimes, a top set is also used on to which the punch is fastened. Cast iron die sets are commercially available which are made complete with guide pins and bushes. Hence the top set can always be located in exactly the same position relative to the bottom set. The press-tool maker then ensures that the punch and die are always in perfect alignment when the tool is closed. This makes the setting operation on the press quick and easy, because the complete tool is simply mounted and fastened on the press, no locating between punch and die being necessary.

2.2 Metal flow

Drawing is a cold-working process in which the operating forces are largely tensile. As the process proceeds, forces must be high enough so that deformation of the work is carried out in the plastic range of the material and therefore the material is in a state of *plastic flow*.

In order to appreciate press-tool design, it may help to briefly reconsider the elementary principles of metal plasticity. Standard tensile or compression tests which cold work the specimen being used are ideal means of obtaining data about the plastic range of metals. Consider Figure 2.2 which shows the results of a tensile test upon a relatively ductile material, such as a low carbon steel.

The metal is elastic up to point *A* and will return to its original size if the force is withdrawn. If the force however, is increased to point *B* before being withdrawn, the force–extension graph follows line *BC*, parallel to line *AO*, as the force is removed. The test piece will then be permanently extended by amount *OC*, and will not return to its former size. *CD* represents the elastic contraction (recovery) which occurs as the force is removed.

Area *OABD* represents the work required to cause deformation *OC*.

If the overstrained material is again subjected to a tensile force upon a testing machine, we shall plot an entirely different force–extension graph than we first derived. This second graph will now have its origin at *C* (instead of *O*), its yield point approximately at *B* (instead of *A*), and its breaking point at approximately the same point as would have occurred if the first test had been completed to failure. In effect, the original piece of metal in being cold worked to point *B* well above the yield point acquires a new set of properties. These new properties result in a different force–extension graph being derived if the metal is reworked. This is the most important first effect of cold working, which means that a cheaper material can be specified for a cold-

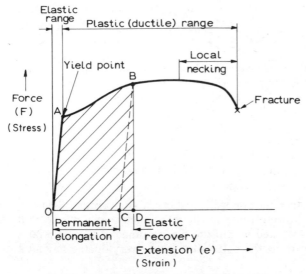

Fig 2.2 Tensile test results of a ductile material

forming operation, the component finishing with new superior properties comparable to a more costly material.

The results of cold working a metal to point *B* well within the plastic range of the metal can be summarised as follows:

(a) The yield point, and hence the stress at yield point is raised, where stress at the yield point is

$$\sigma_y = \frac{\text{Force at yield point}}{\text{Cross sectional area}}$$

(b) The ductility is lowered, and hence the elongation % is reduced, where elongation % = $\dfrac{\text{extension}}{\text{original length}} \times 100$

The process of drawing a simple cup from a plain circular blank puts the material into a state of plastic flow. The displacement of the metal involves some stretching, and hence a certain amount of wall-thinning of the cup always takes place.

2.3 Drawing stresses

Figure 2.3 shows the *stresses* which occur as simple *drawing*, or cupping, takes place. As the blank is crowded into the die with a consequent reduction in diameter, the compressive stress in the undrawn flange will cause wrinkling if this stress exceeds the tensile stress of the material. If, on the other hand, the force on the blank is such that the tensile stress in the wall exceeds the ultimate tensile stress of the

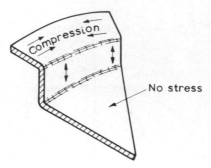

Fig 2.3 Drawing stresses

material, then the cup walls will crack and rupture. Referring to Figure 2.2, the drawing force must be of such a magnitude that the component is worked within the plastic range of the material.

For drawing, the material must be very ductile having a low ratio:

$$\frac{\sigma_y}{\sigma_u}$$

where σ_y is the yield stress and σ_u is the ultimate tensile stress.

Drawing is essentially a straining process leading to surface distortion. The forces involved in the process depend upon the stresses induced by the strain in the material. Figure 2.3 shows a tensile stress in the direction of drawing. A more moderate tensile stress occurs in the flange (combined with the compressive stresses indicated), when a blank holder is in use causing the material to stretch. There is no stress induced, as indicated, in the transverse direction. Hence, the material is mainly under tensile stress causing thinning of the section at most points.

2.4 Stress–strain characteristics

Stress–strain analysis applied to drawing is in fact an uncertain and complicated matter, even with the aid of stress–strain diagrams, and the knowledge required to accurately predict stresses and forces is largely empirical (i.e., based upon experience rather than scientific principles). Figure 2.2 shows that strains in the plastic range associated with plastic deformation are much greater than those in the elastic range. However, these strains vary from stage to stage as the drawing process proceeds, and any changes in thickness are caused by these strains. (See Figure 2.4.)

Alloys are now being developed with the ability to withstand plastic deformation on a large scale without risk of fracture. This phenomenon is known as superplasticity, but alloys so developed to date have limited practical value. However, research shows that metal such

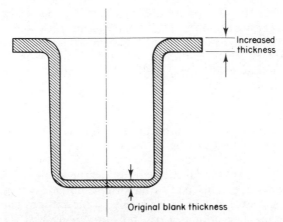

Fig 2.4 Plastic strain in a partially drawn cup

as steel can be rendered superplastic by refining the grain size abnormally small during manufacture. There is no doubt that if technical problems involved are solved, then superplastic alloys will have a profound beneficial effect upon cold-working processes, such as drawing.

Figure 2.5 shows an important relationship between the grain size of the metal after annealing and the amount of cold work applied to the metal before annealing.

This graph shows that there is a critical amount of cold work at which point the grain size of the metal will be coarse. This critical value is shown at point X in Figure 2.5 and is about 10% for a low carbon sheet steel. If a workpiece made from this material receives

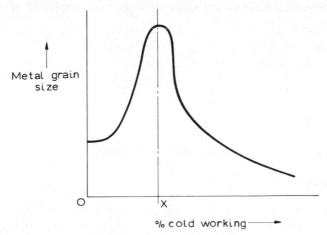

Fig 2.5 Grain size–cold-work relationship

approximately 10% deformation during drawing and is then annealed in order to restore ductility say, a coarse grain will result. This spoils the finish of the article because the large grained structure shows up as an 'orange peel' effect on the surface of the work. Where heat treatment is to follow cold working, the critical amount of cold working should be avoided.

2.5 Blank holder

The *blank holder* operates under pressure on the work blank keeping the metal tight against the die face, and the blank is *ironed out* as it is drawn over the radiused edge of the die hole. This prevents wrinkling of the cup and keeps the finished edge of the rim straight. The pressure which is applied by this pad is important; no pressure gives heavy wrinkling, and excessive pressure results in the bottom being pressed out of the cup. The optimum pad pressure giving best results varies with the type of work, but will be to the order of 30 to 40% of the drawing pressure in the case of circular blanks.

Another aspect of blank holder pressures has arisen due to recent research. This is that blank holding forces considerably in excess of those required to suppress wrinkling improve the efficiency of lubrication existing between the tool and blank faces. This apparently results in deeper draws being possible than would otherwise be the case.

2.6 Draw radius

It is necessary to manufacture both the punch and the die with a *profile radius*, as shown in Figure 2.6, if satisfactory results are to be obtained. The die radius is the most critical as it has been shown that the force required to draw a blank into a cup is virtually independent of the punch radius, but depends upon the magnitude of the die radius.

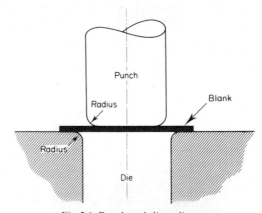

Fig 2.6 Punch and die radius

A generous die radius, but not excessive, gives the best results in reducing the drawing force. The die radius influences the drawing ratio (see Section 2.11), and again it has been shown that a die radius equal to 10 × material thickness allows the maximum drawing ratio to be achieved.

The punch radius is not so critical, but changes of thickness which occur during drawing are influenced by the punch profile. A generous punch radius is desirable as it maintains the thickness, and hence the strength, of the cup walls.

2.7 Die cushions

Figure 2.1 shows a *pressure pad* acting in the die aperture supporting the cup as it is drawn into the die. Hence, puckering of the base of the cup is prevented as it is firmly gripped between the faces of the punch and pressure pad. Figure 2.7 shows a bending tool for producing a

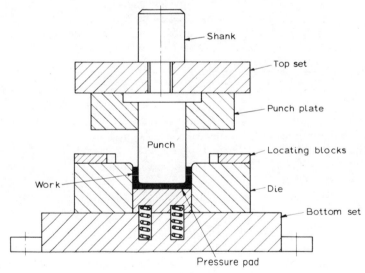

Fig 2.7 Bending tool

simple U-shaped component. This tool similarly contains a pressure pad which is performing a similar function.

In both cases the pressure pad can also act as an ejector to push the component clear of the die. In certain instances depending upon the component characteristics, the completed workpiece will stick to the punch instead of the die. In such a case a means must be provided for stripping the component off the punch.

The disadvantage of spring actuated pressure pads is that pressure increases with extended punch travel, i.e., each increment of travel increases the pressure upon the pad. A better alternative is provided by pneumatic means through the use of air cushions. Such a *die cushion* is fitted below the die set, and is usually attached to the underside of the press frame as shown in Figure 2.8.

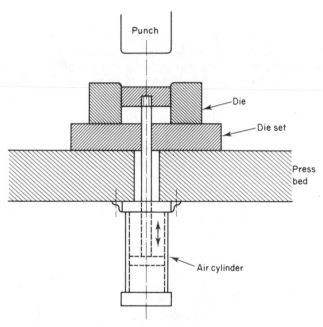

Fig 2.8 Die cushion

Reference to Figure 2.8 will show that the air in the cylinder is compressed on the down stroke. This air is forced to flow into a tank. On the return stroke the air in the tank forces the piston back to its original position. The advantage of die cushions is that the die pad pressure remains constant over the full punch travel, and also the pressure may be varied. In addition to pneumatic die cushions, hydraulic and hydropneumatic die cushions are also used for heavier applications.

Figure 2.9 shows an alternative form of bending tool where the pressure pad firmly holds the workpiece while bending takes place under the action of the punch.

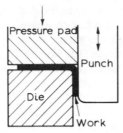

Fig 2.9 Bending tool

2.8 Redrawing

Where it is impossible to draw a long cup to the required depth in one operation, it will be necessary to sub-divide the process into a series of draws. Hence, in addition to a drawing tool, one or more *redrawing tools* will be required. Figure 2.10 shows a typical redrawing tool ar-

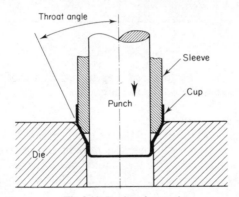

Fig 2.10 Re-drawing tool

rangement using the direct method. The blank holding arrangement shown here is different from that on the tool at Figure 2.1. The first-stage cup only has to be centred in the throat of the die initially, then the descending punch positively locates the cup in the die and completes the draw. The concentric sleeve shown fitted to the punch assists the initial centring and location. A throat angle of about 15° has been shown to be satisfactory for most purposes. Figure 2.11 shows a re-drawing tool using the reverse method. In this case the first-stage cup is located upside-down on the outside of the die. The descending punch then turns the cup inside-out; hence the amount of cold working is increased. This results in an improved material structure, and there is less likelihood of a coarse grain developing due to the critical amount of cold work having been applied. (See Section 2.4)

It is sometimes possible to complete all the necessary redrawing operations without intermediate annealing being required, but the total reduction must then be made within the plastic range of the material as indicated by the elongation %. (See Section 2.2) Generally however, where several redraws are necessary, sufficient annealing

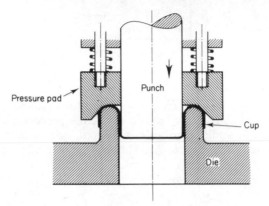

Fig 2.11 Re-drawing tool

must be carried out between redraws to relieve the work-hardening induced by the excessive cold working.

2.9 Multi-stage drawing

Where cups which are long in relation to their diameter are required to be produced by a drawing process, it may be impossible to effect the operation in a single draw. Hence, one or more redrawing operations will be required (as described in Section 2.8) and the process becomes a *multi-stage* one. This will be so where the cup material ratio $\frac{\sigma_y}{\sigma_u}$ is too high, and the punch force necessary to draw the cup in one operation causes the material to rupture. That is, the induced tensile stress exceeds the material ultimate tensile stress.

Although the reasons are not absolutely clear, the amount of reduction possible in redrawing is less than that which can be obtained by drawing from a flat blank. Figure 2.12 shows the degree of redrawing which is usually possible using a ductile material. In the example shown a nominal 100 mm diameter blank is drawn in stages to a 33 mm diameter cup.

2.10 Blank size

In order to calculate the *blank diameter* 'D' to produce a cup of diameter 'd', it is necessary to equate the surface area of each, on the

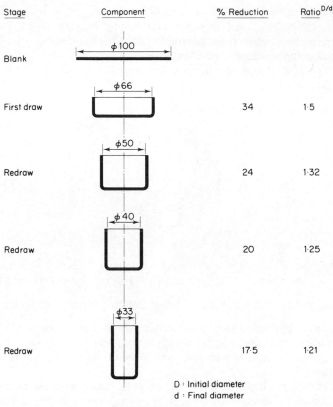

Stage	Component	% Reduction	Ratio$^{D/d}$
Blank	φ100		
First draw	φ66	34	1·5
Redraw	φ50	24	1·32
Redraw	φ40	20	1·25
Redraw	φ33	17·5	1·21

D : Initial diameter
d : Final diameter

Fig 2.12 Drawing stages

assumption that no change in metal thickness occurs and that there is no corner radius at the bottom of the cup.

$$\text{Surface area of blank} = \frac{\pi D^2}{4}$$

$$\text{Surface area of cup} = \frac{\pi d^2}{4} + \pi dh$$

$$\therefore \qquad \frac{\pi D^2}{4} = \frac{\pi d^2}{4} + \pi dh$$

$$\therefore \qquad D = \sqrt{d^2 + 4dh}$$

where h = cup height.

In practice the above assumptions are not correct, so a precise value for 'D' might have to be obtained by trial and error.

Example 2.1

A cup is to be drawn to a diameter of 76·2 mm and a height of 38·0 mm. Estimate the blank diameter.

Solution

$$
\begin{aligned}
\text{Blank diameter } D &= \sqrt{d^2 + 4dh} \\
&= \sqrt{5800 + (4 \times 76\cdot2 \times 38\cdot0)} \\
&= \sqrt{5800 + 11\,582\cdot4} \\
&= \sqrt{17\,382\cdot4} \\
&= 131\cdot84 \text{ mm}
\end{aligned}
$$

2.11 Drawing ratio

The ratio $\dfrac{D}{d}$ where D is the blank diameter and d is the cup (punch) diameter, is known as the *drawing ratio*. It is a useful measure for expressing the severity of a drawing operation. It has been found that $\dfrac{D}{d}$ should not exceed a value of 2 when drawing a material under good conditions, although this will be influenced by the material thickness and its ultimate tensile strength. Figure 2.11 shows values of drawing ratios for drawing operations which represent typical industrial practice.

The maximum drawing ratio that a metal can reach gives a good indication of its suitability for deep drawing. However, the success of deep drawing is dependent upon the quality of the material, and this quality is best measured by a cup drawing test of the Erichsen type.

2.12 Drawing force

The precise value of a *drawing force* is difficult to accurately estimate because the force varies as the operation proceeds, and the punch force will decrease as the blank diameter decreases (and the depth of draw increases). However, the drawing force increases proportionally to the drawing ratio, i.e., the greater the initial blank diameter for a given cup size, so proportionally greater will be the drawing force.

Empirical formulae have been established to enable the maximum drawing force to be estimated, and the following expression will give a sensible value:

$$
F_{\max} = \pi \cdot d \cdot t \cdot \sigma_u
$$

where d = cup diameter (internal)
$\quad\ \ t$ = material thickness
$\quad\ \ \sigma_u$ = material ultimate tensile strength.

Example 2.2

A cup is to be drawn to a diameter of 76·2 mm from 0·8 mm thick material. Estimate the maximum drawing force required to press the cup if $\sigma_u = 432$ N/mm^2

Solution

$$F_{max} = \pi . d . t . \sigma_u$$
$$= 3·142 \times 76·2 \times 0·8 \times 0·432 \text{ MN}$$
$$= 0·083 \text{ MN}$$

Learning Objectives 2
The learning objectives of this section of the TEC unit are:

2. Analyses and applies principles of metal forming.

2.1 Explains the functions of the elements of a drawing tool.
2.2 Describes the flow of metal in a typical drawing operation.
2.3 Explains the nature of stresses induced, with reference to 2.2.
2.4 Examines stress–strain characteristics and other data and relates to the drawing process.
2.5 Describes the effect of varying blank holder pressure.
2.6 Describes the effect of draw radius.
2.7 Explains the use of die cushions in press tools for bending and drawing operations.
2.8 Describes the technique of redrawing.
2.9 Identifies the conditions which make drawing in two or more stages desirable or essential.
2.10 Calculates blank size for a given cup.
2.11 Explains the significance of the drawing ratio.
2.12 Calculates the maximum force required for drawing.

Exercises 2

1   Sketch a typical drawing tool and identify the major elements of its construction. Briefly describe the function of each element.
2   Describe the state of plastic flow associated with a simple metal drawing operation, and comment upon the stress–strain characteristics.
3   Describe the effect of a varying blank-holder pressure when carrying out a drawing operation.
4   What is the effect of a die draw radius, and a punch draw radius respectively on a drawing operation?
5   What are die cushions, and what is their purpose?
6   Sketch a typical redrawing tool. Describe the conditions which make the use of such a tool essential.

7 What is meant by multi-stage drawing, and when would the technique be used?

8 (*a*) Derive an expression for estimating the blank diameter '*D*' of a cup of diameter '*d*' × height '*h*'.
   (*b*) Estimate the blank diameter for a cup which is to be drawn to 62·3 mm diameter × 21·4 mm height. (Ans. 96 mm.)

9 What is meant by the term 'drawing ratio'?

10 Estimate the maximum force required for drawing a cup of 62·3 mm internal diameter in 0·8 mm thick material if $\sigma_u = 325$ N/mm². (Ans. 0·051 MN.)

Chapter 3
Powdered Materials

In recent years the process known generally as 'powder metallurgy' has developed to such an extent that it is now recognised as an important method of producing metal shapes. The utilisation of metal powders is increasing at the rate of about 15% per year as new variations of the process are exploited. For example, it seems that the production of stainless steel strip from powder is economically viable. The process is divided into four main stages:

1. Production and mixing of the powders. Powders may be produced in a variety of ways including grinding or pulverising in a suitable mill, chemical reduction (such as reducing the metal oxide), atomisation or electrolysis.
2. Compacting or pressing the powder(s) in a suitably shaped die.
3. Sintering the resulting 'green' compact in a furnace.
4. Sizing where required.

The cost of producing powders is high. Compacting dies and sizing dies are expensive. The time taken to sinter a green compact is high. Hence, powder metallurgy is not an economic process unless large production runs of components are possible.

3.1 Powder selection

Metals, and other materials such as the polymers or alumina, can be powdered. The process of powder metallurgy is applied to most of the common metals with the exception of aluminium which has an affinity for steel, hence causing difficulties at the pressing stage, by its tendency to cold-weld itself to the die. The *selection of powder*, or powders, will depend upon the resulting structure that is required.

In the case of refractory metals like tungsten which have a very high melting temperature, it is, for example, very costly to melt them in order to cast them into an ingot. It has been found that the cost of sintering powdered refractory metals in compacted form is cheaper. Hence, in such a case a single powdered metal is used.

Where an alloy of two or more materials is required, powdered metals of say 0·002 mm (2 microns) particle size are thoroughly mixed together, then compacted and sintered. Certain metallurgical structures can only be achieved by this means, for example, the structure of a 2-phase system such as Cobalt–Tungsten Carbide as shown at Figure 3.1(*b*).

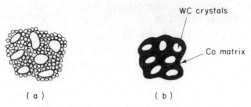

Fig 3.1 Co–WC structure

Figure 3.1(*a*) shows the original mixture of powders, and Figure 3.1(*b*) shows the resultant structure in which the hard tungsten carbide (WC) crystals are dispersed in a tough, but ductile, matrix of cobalt (Co).

An alloy-type mix of two powdered metals can be achieved where the metals are completely insoluble in both the liquid and solid state (such as Zinc and Lead), or partially insoluble in both the liquid and solid state (such as Iron and Copper). If two such metals are melted and cast together they will not give a satisfactory structure as they will separate out into individual layers. However, mixing together in the required quantities followed by sintering will cause dispersion of the two metals, in effect giving an alloy of the required type.

An example of this is illustrated at Figure 3.2, where Figure 3.2(*a*) shows the original mixture of powders, and Figure 3.2(*b*) shows the resultant structure in which the different powdered materials have completely diffused into each other.

Fig 3.2 Diffused metal powder structure

Given the choice of structures described above (and their variants), the factors influencing the selection of the powder are:

(*a*) Type of powder(s) and proportions;

(*b*) Powder sizes in the mixture.

Consider each factor in turn:

(*a*) Powdered materials can now be produced having particles of varying shapes. The simple spherical shaped particle has the disadvantage of poor compressibility because a sphere has the minimum surface area for a given volume; but less pressure is required for a given density as round particles slide easily over each other. Spherical powder is preferred for the production of filters, because the relatively looser packing of the powder in the mould leads to the desired finished structure.

In the case of a powdered metal product to be used as a cutting tool, such as a cemented carbide tip, angular shaped particles are to be preferred dispersed throughout a softer matrix (see Figure 3.1.). In effect a series of multi-point, minute cutting tools are available.

Take one more example; that of components requiring magnetic characteristics. Here the powder particles making up the magnetic structure will give best results if they are needle-like in shape.

(*b*) The powder size influences such things as density of the compact, consolidation time and the uniformity of the finished structure. Higher compact density is more likely to be achieved using say two particle sizes rather than particles of a similar size. Figure 3.3 shows such a mix

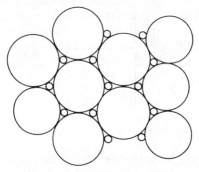

Fig 3.3 Structure of different particle size

where, for example, the particle sizes might be approximately 1 micron and 10 microns respectively, Section 3.6 shows that the sintering process can markedly affect the porosity of the finished product, but powder size may also influence this characteristic.

3.2 Die design
The accuracy and finish of the green compacts pressed to shape in a *die* depend entirely upon the degree of accuracy and finish to which the die is manufactured. The process is an abrasive one and the frictional conditions are severe, so very hard die metal must be used such as tool steel or tungsten carbide. In addition a die lubricant is usually used, such as oil or graphite solution, or a metal stearate may be mixed with the powder.

Die pressures may rise as high as 1500 N/mm², especially when compacting very hard metals, so dies are manufactured as solid entities rather than in sections. Hence, components cannot be extracted from the die by having it open for component removal, so the solid die must therefore be designed to allow extraction.

Powdered metal does not behave as fluid molten metal would when under pressure in a mould cavity, and there is no uniform pressure distribution. Powder will not freely flow laterally (sideways) in the die and this must be taken into account at the design stage. It is necessary to apply pressure in the die using both an upper and lower punch in order to give uniform pressure distribution. This is shown at Figure 3.4.

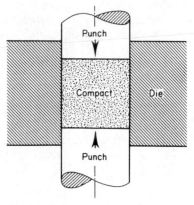

Fig 3.4 Uniform pressure compacting

Large capacity presses of either mechanical or hydraulic operation are used, and the means must be provided of automatically feeding the precise amount of powder into the die. All means of high production must be exploited, and where the capacity allows multi-cavity dies should be used. Automatic, fast, large volume production is required if the process is to be economically viable.

3.3 Powder compaction

Take the example of a simple bush which is to be produced in quantity by the powder metallurgy process. The external shape and size will be controlled by the die cavity, and the bore shape and size will be controlled by a core rod. There will be a big reduction in volume at the *compacting stage*, and the compression ratio varies between 2/1 and 3/1. Further, after the sintering stage there will be an overall reduction of the compact volume of up to 20%, and this must be allowed for in the die size. Figure 3.5 shows the press operation cycle for a die designed to produce a simple bush.

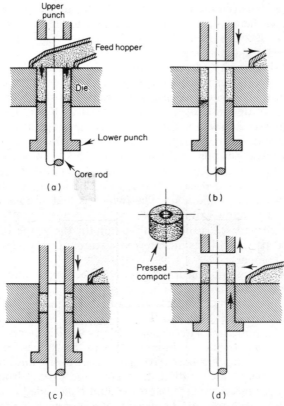

Fig 3.5 Compaction of powdered metal

The cycle of the compacting operation shown is as follows:
(*a*) the feed hopper covers the die and the cavity is automatically filled with the metal powder mixture;
(*b*) the feed hopper moves clear of the die cavity as the upper punch descends;
(*c*) the upper and lower punches move vertically together to press the powder into a compact of uniform density;
(*d*) the upper punch returns, and the lower punch makes a further vertical upwards move to eject the compact from the die. The feed hopper returns and the lower punch returns to complete the cycle.

The green compacts will now be conveyor fed into a sintering furnace. At this stage the compacts have sufficient mechanical strength to be handled without breaking. This is possible because a 'cold-weld' has been achieved between the powder particles due to the severe pressure imposed upon them.

3.4 Shape limitations

Section 3.2 described some of the important factors which must be taken into account at the die design stage. Clearly, the limitations on die design will affect possible *component shapes*. Also it should be remembered that simple shapes can be pressed at higher speeds than complex shapes. Below are listed some component features which should be avoided in the shape of components which are to be produced from metal powder.

(*a*) Re-entrant profiles, such as dovetails, and reverse tapers. Compacts having these features would be impossible to extract from a solid die, and a split die (which is undesirable) would be required to allow extraction.

(*b*) Sharp corners. It is difficult to maintain the shape of a sharp corner, particularly as an external feature, upon a relatively soft green compact.

(*c*) Thin sections and abrupt changes of section. A chunky, solid object of reasonable size is easier to press and handle than delicate, thin-sectioned objects of elaborate shape. Feather edge features are also difficult to retain.

(*d*) Complex holes and lateral holes. Plain, vertical bores as shown in the component at Figure 3.5 can be successfully cored. However, a complex bore having a change of size and shape, or a lateral hole would be difficult to core and extract from a solid die. Loose pieces and split cores should be avoided.

(*e*) Length-to-width ratio. The density of a compact at the pressing stage decreases over its length. Although double-ended compaction (as shown at Figure 3.4) helps to give uniform pressure, and hence density, it is still possible with extra long components to get a region of low density in the centre. Hence ratio: $\dfrac{length}{width}$ greater than 3:1 is not recommended.

3.5 Sintered products

For convenience, *sintered products* can be grouped into four categories as follows:

(i) Refractory material products. These are parts manufactured from refractory materials having very high melting temperatures such as Tungsten which melts at 3400°C. Many of these are now cheaper pressed from powder and sintered, rather than melting and casting to shape in a mould. In any event, the latter alternative is not a practical proposition.

Products fitting into this category for example are cutting tool tips made from tungsten carbide, or alumina (aluminium oxide).

(ii) Porous material products. By varying the volume of the powder charge and the compacting pressure it is possible to control the degree of porosity in the structure of the finished component. A highly porous structure having a porosity above 20% gives a useful bearing characteristic to the material as the pores can be charged with a lubricant. Such porous (or permeable) structures also have useful filtering properties.

Products fitting into this category for example are bronze bearings or stainless steel filters.

(iii) Composite material products. Using a mixture of two or more powders it is possible to manufacture parts with a structure which otherwise would be impossible to achieve. Such 'alloys' have properties which make them commercially useful, and they are used extensively for electrical products. It is true to say that the whole field of composite sintered products is in its infancy.

An example of a composite material sintered product is an electrical current collector brush.

(iv) Standard structural products. Although the cost of metal powders is still comparatively high, it is expected that the cost of sintered products will become more and more competitive as high production techniques are exploited. Many steel products, particularly of difficult and precise contour, are now produced from powdered metals as opposed to machining from solid metal. Sintering can also give a marked increase in electrical conductivity which can be utilised in certain electrical products.

Examples of such products are cams, precision gears, small permanent iron magnets.

3.6 Sintering

Sintering is heat treatment carried out on the green compact at an elevated temperature which is below the melting temperature of the particular metal or 'alloy'. Sintering temperatures are usually between 60 and 90% of melting temperatures. In a two-phase system, one element of the powder mixture will often be in a liquid state. In such a case it is thought that the melted metal is drawn by capillary action to fill the spaces between the particles of the unmelted metal.

The components are fed through a gas, electric or oil-fired furnace on a metal belt at a speed sufficient to complete the sintering process. It is a relatively slow process, measured in hours rather than minutes, and the sintering time affects the resultant structure. Sintering is carried out in a protective atmosphere in order to prevent oxidisation. Carbides are usually sintered in a hydrogen atmosphere; for example silicon carbide at a temperature of 1350°C. Hydrogen is costly, and often dissociated ammonia containing 75% hydrogen and 25% nitrogen is used as a cheaper alternative, except where an adverse reaction with nitrogen might occur. The cheapest, and most widely used atmospheres, are those produced by partial combustion of hydrocarbons. For example, a hydro-carbon gas such as methane cracked under heat with insufficient air for complete combustion, gives a protective atmosphere containing up to 45% hydrogen.

Sintering is basically a welding process in the initial stage in which the cold-welding which took place at the compacting stage is consolidated by 'weld-necks' being established between the particles. This is shown at Figure 3.6.

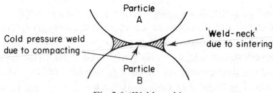

Fig 3.6 'Weld-neck'

In these initial stages of sintering as the compacts are soaked in the heat, no significant change in density occurs, although there is a marked increase in the electrical conductivity of the sintered part. The porosity can be controlled at this stage, and porous bearings would receive no further sintering.

In the final stages the pores begin to close, and there is a big change in the permeability, and hence density, of the structure. The original volume of the compact may have decreased by up to 20% if it is left in the furnace long enough to pass through all the sintering stages. The particles which were cold-welded together through pressing, are closely knitted together in the final structure due to volume diffusion. This bonding together of the particles due to sintering gives a considerable mechanical strength, comparable to that of alloys, produced by melting.

3.7 Advantages and limitations
The *advantages* and *limitations* of sintered products may be summarised as follows:

Advantages
    (i) certain metallurgical structures can only be produced using the sintering process applied to powdered materials;
    (ii) certain of the very hard materials and compounds can only be produced at reasonable cost by pressing and sintering, as opposed to melting and casting;

(iii) certain sintered products, such as magnets, often have a superior quality to those produced by more orthodox means;

(iv) sintered products can be produced to a high degree of accuracy and finish, requiring little or no machining with the consequent elimination of swarf;

(v) many sintered parts can now be produced more quickly, and hence more cheaply, than equivalent parts produced by more orthodox means;

(vi) a saving in space as powdered metal is more convenient to store than bar stock.

Limitations

(i) the cost of compacting dies and sizing dies is high;

(ii) dies wear quickly because of the high pressures used and the frictional conditions of the process;

(iii) powdered materials are relatively costly to produce, although the cost is decreasing as the usage of the process increases;

(iv) the cost of sintering is relatively high;

(v) high production runs are essential in order to make the process economically viable;

(vi) there is a limit to the size of the sintered part which can successfully be compacted at the high pressures required;

(vii) certain difficult contours cannot be produced in closed compacting dies.

3.8 Post-sintering treatment

Sintering is a heat-treatment process carried out at a relatively high level although below melting temperatures. Inevitably, therefore, some distortion takes place with minor changes to size and shape. If the changes are significant then a *sizing operation* will be necessary after sintering in order to restore the final shape to its required dimensional accuracy and surface finish.

Sizing is in effect a re-pressing operation in a die similar to that used for compacting, and in fact the press tools are basically the same in principle in that punches and dies of the appropriate contour are required. Mechanical or hydraulic presses are used, and the process is a high production one. The following figures give an indication of the dimensional improvement that takes place as a result of sizing:

Average compacting tolerance: 0·05 mm/25 mm length.

Average sizing tolerance: 0·02 mm/25 mm length.

These tolerances are typical for dimensions which are perpendicular to the direction of pressing; for those parallel to the direction of pressing the tolerances would probably be twice the stated magnitude.

Learning Objectives 3
The learning objectives of this section of the TEC unit are:

3. Evaluates the process of and applies the principles of manufacturing components from powdered material.

3.1 States the factors influencing correct selection of powder.
3.2 States the factors influencing die design.
3.3 Analyses, for a given component, the effects of die design and press operation on powder compaction.
3.4 Relates die design limitations to component shape.
3.5 Lists typical examples of sintered products.
3.6 Describes the sintering process.
3.7 Appraises the capabilities and limitations associated with sintered products.
3.8 Justifies the application of post-sintering treatment.

Exercises 3

1 State the factors which influence the correct selection of powder for a powder metallurgy process.
2 With the aid of sketches describe the factors influencing the design of a powdered metal die.
3 Sketch a simple component which would be suitable for production by the powder metallurgy process. Sketch the die for producing such a shape, and indicate the method of operation.
4 Name six typical products of different types which are produced by the powder metallurgy process.
5 Describe sintering referring to the different stages of the process.
6 List and briefly expand upon the advantages and limitations of the powder metallurgy process.
7 A component can be produced by the alternative processes of powder metallurgy or forging. What factors would be taken into account in order to justify the choice of process?

Chapter 4
Centreless Grinding

The external version of this process is now extensively used for work which was previously ground between centres on a plain grinding machine. It is a commonplace quantity grinding process for parallel, tapered or formed work which is required finished to a high degree of accuracy. External centreless grinding can, for example, produce ground bar stock, such as 'silver steel', to a high degree of accuracy in large quantities at a competitive price.

There is an internal version of this process, which is much less common but offers certain advantages over plain internal grinding.

However, the term 'centreless grinding' is usually applied only to the external process which is described in this chapter.

At the roughing stage, cuts from 0·03 mm to 0·25 mm are usual. Finishing cuts can achieve a dimensional accuracy to the order of 0·005 mm.

4.1 Centreless grinding machine
Figure 4.1 shows the end view of a typical *centreless grinding machine*. Figure 4.2 shows the principle of operation of the machine.

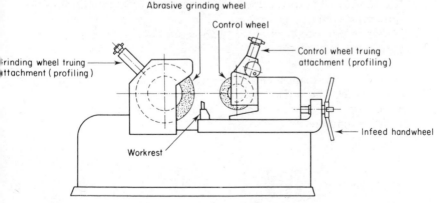

Fig 4.1 Centreless grinding machine

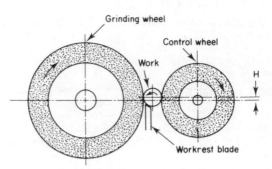

Fig 4.2 External centreless grinding

The system consists of three elements:

1. *Grinding wheel.* This is the normal type of plain, straight wheel which grinds on the outside face as used for the more orthodox cylindrical grinding of work between centres. It rotates at a peripheral speed of 25–30 m/s. As viewed in Figure 4.2 it rotates in a clockwise direction throwing the stream of sparks downwards. In diameter it varies from about 300 to 500 mm, and the width, grit and grade of the wheel are chosen to suit the type of work and operation.

2. *Control wheel.* It is on the same horizontal centre-line as the grinding wheel. Again it is an orthodox, straight abrasive wheel but of rubber bond. It regulates the rotary action of the work, driving it as if they were two friction wheels. The peripheral speed of the work will be that of the control wheel, if there is no slip. The control wheel rotates in the same direction as the grinding wheel, hence the work rotates in an anticlockwise direction as viewed in Figure 4.2.

Control wheels vary in diameter from about 200 to 300 mm, and the speed range available on the machine would be say 12 to 200 rev/minute.

Example 4.1

Work of 34 mm diameter is to be ground using a 250 mm diameter control wheel. The surface speed of the work is 0·4 m/s.

Calculate the work speed and control wheel speed in revolutions/minute.

Solution

$$\text{Work speed } n = \frac{0\cdot4 \times 1000}{34\ \pi} = 3\cdot74 \text{ rev/s}$$
$$= 224\cdot4 \text{ rev/min.}$$

The work peripheral speed will be that of the control wheel, hence:

Control wheel speed $N = 224\cdot4 \times \dfrac{34}{250} = 30\cdot52$ rev/min.

$$\text{say } 30 \text{ rev/min.}$$

3. *Work rest*. This incorporates work guides for through-feed grinding, and a work blade which has a 30° angular top face. The blade is often stellite faced to resist wear. The angular top face is necessary for three reasons:

(*a*) to keep the work in close contact with the control wheel face;

(*b*) to assist in the 'rounding-up' process;

(*c*) with in-feed grinding, to allow the workpiece to gradually be brought into contact with the grinding wheel.

The workrest blade is set at such a height that the work axis is positioned at a suitable height above the wheel centres as shown at Figure 4.2. This height 'H' should not be less that $\frac{1}{8}d$ or greater than $\frac{1}{2}d$, where 'd' is the work diameter, but $H = \sqrt{1\cdot6d}$ gives a satisfactory value.

Example 4.2

Calculate the height at which the work centre should be set above the wheel centre if the work diameter is 34 mm.

Solution

$$H = \sqrt{1\cdot6d} = \sqrt{1\cdot6 \times 34}$$
$$= \sqrt{54\cdot4} = 7\cdot376$$
$$\text{say } 7\cdot4 \text{ mm}$$

When small diameter work is being ground it is usual to set the workrest as close as possible to the abrasive wheel, in order to prevent the likelihood of the workpiece falling between the blade and the abrasive wheel.

The work-guides are necessary to ensure that the work lines up accurately with the wheel faces as it feeds past them, as shown at Figure 4.3, which is a plan view. Figure 4.3 indicates the limits of clearance between the guides and the work. The clearance on the guides on the control-wheel side is particularly important if straight work is to be produced. If the control-wheel guide is set too far forward then the work will be ground with a barrel effect, and if set too far back then the work will have a bell effect.

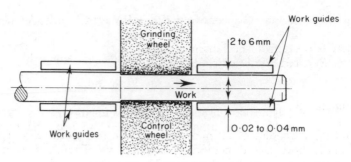

Fig 4.3 Work guides

4.2 Operating methods

There are *three variations* of the process:

(*a*) *Through feed*. Used for parallel work of any length which has no surface obstructions such that it can be passed completely between the wheels set at a fixed distance apart to give the correct work diameter.

The principle of through-feed grinding, or straight-through grinding as it is sometimes called, is shown at Figure 4.4.

As can be seen from this diagram, the control wheel is tilted through angle θ in order to impart axial feed to the work. From the velocity diagram at Figure 4.4 the work speed $= W \sin \theta$

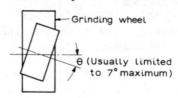

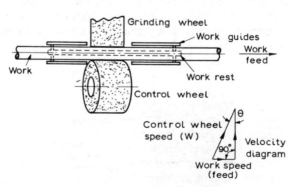

Fig 4.4 Through feed grinding

$$\therefore \qquad \text{the work-feed } F = \pi DN \sin \theta$$

where D = control wheel diameter (mm)
$\quad\ \ N$ = control wheel speed (rev/s)

Example 4.3

Plain steel spindles of 30 mm diameter are to be centreless ground using the through-feed method. The control wheel which is set at 1° to the work axis is 250 mm diameter. The peripheral work speed is 0·20 m/s. Assuming no slip between the work and control wheel, calculate (a) work speed n; (b) control wheel speed N; (c) work speed F in mm/s.

Solution

$$\text{Work speed } n = \frac{0\cdot20 \times 1000}{30} = 2\cdot12 \text{ rev/s}$$

$$\text{Hence, with no slip, } N = 2\cdot12 \times \frac{30}{250} = 0\cdot25 \text{ rev/s}$$

$$\begin{aligned}
\text{Work feed } F &= \pi DN \text{Sin } \theta \\
&= 3\cdot142 \times 250 \times 0\cdot25 \times \sin 1° \\
&= 3\cdot142 \times 250 \times 0\cdot25 \times 0\cdot0175 \\
&= 3\cdot44 \text{ mm/s}
\end{aligned}$$

Through feed is the simplest method used, and is ideal for the production grinding of plain parallel parts, such as rollers, which can be hopper fed. Long bars can also easily be ground at speed by this method, with the process virtually being continuous.

(b) *Infeed.* This is plunge grinding and is used for headed work, or multi-diameter work for example, which cannot be passed axially straight through the wheels. The principle of infeed grinding is shown at Figure 4.5.

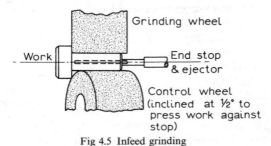

Fig 4.5 Infeed grinding

The work is placed against a pre-set end stop and the control wheel slide advances the rotating work up to the grinding wheel. When the slide meets a stop the grinding wheel 'sparks out' to leave the job at the correct diameter. The slide (and hence work) retracts and the work is then automatically ejected by the end stop.

This cycle may be operator controlled but again lends itself to automatic control. This is usually done by hydraulic means. Figure 4.6 shows the timing cycle.

When work is to have a formed profile ground upon it, then it is necessary to dress the grinding wheel to the correct form using a former plate. A follower, which is connected to the diamond dresser,

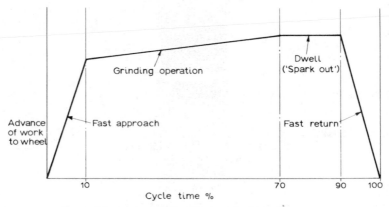

Fig 4.6 Automatic cycle for infeed grinding

traverses along the profile on the former plate and in turn produces the required form on the grinding wheel. Similarly, the control wheel is dressed, but a full form is not necessary as this wheel does not cut. This dressing operation is illustrated at Figure 4.7.

(c) *Endfeed.* This is sometimes called the *plunge and run* technique and may be used for headed work having a shank length which is too long for the wheel face, hence preventing the use of infeed grinding. In practice, it is almost always used for taper grinding.

The principle of endfeed grinding is shown at Figure 4.8.

This is in effect a process which is a mixture of through feed and infeed. The work is plunge ground to size firstly, followed secondly by a through feed to an end stop which grinds the whole shank diameter to size. Alternatively simple endfeed grinding can be carried out for taper work without any plunge grinding being done.

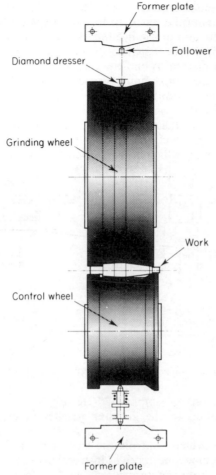

Fig 4.7 Wheel dressing

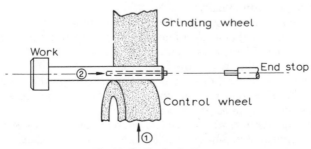

Fig 4.8 Endfeed grinding

4.3 Choice of Method

Three different components will be considered here having different characteristics. The first is a gudgeon pin which requires to be ground on the outside surface to a high degree of accuracy and parallelism. The through-feed method is chosen, and because the components are required in large quantities at an economic price, a fully automatic grinding and sizing set-up is used whereby the parts are 100% automatically inspected after grinding.

Figure 4.9 illustrates the process. Note that the work guides are not

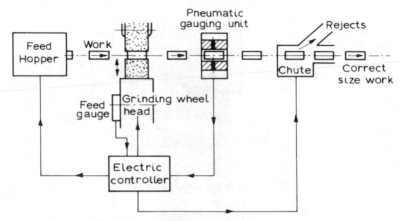

Fig 4.9 Automatic centreless grinding operation

shown in the interests of simplicity. Referring to Figure 4.9 it can be seen that in addition to the grinding machine, there are five other elements:

1. Feed hopper which controls the feed of the components to the grinding machine, and receives instructions in the form of electrical impulses from the controller. The feed hopper need only be filled with components from time to time, and the system will operate automatically.
2. Pneumatic gauging unit which contains pre-set inspection air jets which sense the work diameter, and send out impulses to the controller if the work diameter increases beyond the top limit, due to grinding wheel wear.
3. Electric controller which receives impulses from the gauging unit, and sends out appropriate instructions to the grinding wheel head feed mechanism to feed in.
4. Feed gauge (or wheel head slide positioning unit) which feeds back impulses to the controller so that the wheel slide is set to the correct position.

5. Reject chute which normally allows work through to the finished work hopper when the system is stable. When corrective measures are being taken to readjust the wheel slide position due to impulses received by the controller, instructions are sent to the chute which operates a deflector to divert work to a reject box. The small number of possible rejects involved can be occasionally collected and gauged by an inspector.

The second component to be considered is a spindle bearing 'cone' which requires the outside surface to be accurately form ground. The set-up is illustrated at Figure 4.10 where the infeed method is used.

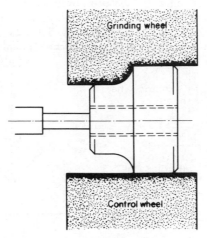

Fig 4.10 Form grinding

The third component is a morse taper sleeve which must be externally taper ground to a high degree of accuracy. The endfeed method is appropriate for this operation which is shown at Figure 4.11. The endfeed method is used extensively for taper grinding, and it may be

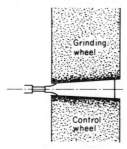

Fig 4.11 Taper grinding

noted that both wheel faces can be dressed to half the taper included angle (as shown at Figure 4.11), or only the grinding wheel face may be dressed to the taper included angle.

4.4 Lobing

There is always a tendency for centreless grinding, where the work is not generated between the location of fixed axial centres, to produce *lobed* work. This is the term used to indicate work which is in cross-section of a constant diameter shape instead of a truly geometric circular shape. Figure 4.12(*a*) shows a lobed, or constant-diameter figure, based upon 120° spacing. If measured across the diametral positions shown it would appear to be circular. However, if rotated in a vee-block underneath a dial test indicator (D.T.I.), as shown at Figure 4.12(*b*), the out-of-roundness defect would be shown.

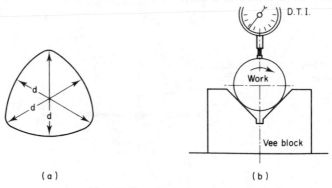

(a)   (b)

Fig 4.12 Out-of-roundness testing

It is very difficult under production grinding conditions, where speed as well as accuracy is demanded, to produce perfectly cylindrical work when the original bar stock itself is not round as is usually the case. If such a bar is fed through the wheels of a centreless grinding machine, the act of grinding does not necessarily correct the original defect. In Section 4.1 it was stated that the angular top face of the work rest blade was necessary to assist in the rounding-up process. In the next section it will be shown that in addition the settings are also important.

4.5 Machine setting

Consider the case of a steel bar which has a high spot along its length, and is being centreless ground by the through-feed method. Also assume that the nominal work axis is at the same height as the wheel centres, and that the work blade top is flat instead of angular. As the work rotates, the high spot will come into contact with the control wheel face. The work will deflect towards the grinding wheel face hence causing an opposite flat spot to be ground. Further, during work rotation the high spot will contact the work blade face hence causing the work to lift. Consequently, the nominal work axis will be raised. The situation is shown at Figure 4.13 where the high spot is obviously exaggerated.

It has been found that as the grinding process continues under these conditions, a tri-lobed work section is produced of the type shown at Figure 4.12(*a*).

A solution to the problem is found by having an angular faced work blade, and by adjusting it such that the work axis is raised above the wheel centres as shown at Figure 4.2. These *settings* will result in the three work contact points being at random angles to each other, rather than at 90° to each other which is the result of settings used at Figure 4.13, which in turn lead to lobing.

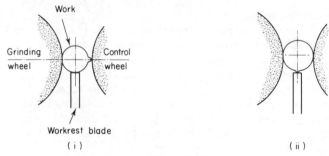

Fig 4.13 Production of lobed work

4.6 Through-put times

Consider the following example which shows how a *time estimate* for centreless grinding may be arrived at:

Example 4.4

Hardened steel spindles of 25 mm diameter by 130 mm long are to be ground by the through-feed method of centreless grinding. The machine has a grinding wheel 150 mm wide, and the control wheel is 250 mm diameter set at an angle of 3°. Assuming a peripheral work speed of 0·20 m/s without slip between the work and control wheel, calculate the cycle time per piece in seconds.

Solution

$$\text{Control wheel speed} = \frac{0\cdot2 \times 1000}{250\ \pi} = 0\cdot25 \text{ rev/s}$$

Work feed $F = \pi DN \sin \theta$
$$= 3 \cdot 142 \times 250 \times 0 \cdot 25 \times \sin 3°$$
$$= 3 \cdot 142 \times 250 \times 0 \cdot 25 \times 0 \cdot 0523$$
$$= 10 \cdot 27 \text{ mm/s}$$

The first component through will have a traversing length of 130 + 150 = 280 mm

$\therefore$ Cycle time $= \dfrac{280}{10 \cdot 27} = 27 \cdot 26$ seconds.

Subsequent components feeding through 'bumper-to-bumper' will have an effective traversing length of 130 mm.

$\therefore$ Cycle time/piece $= \dfrac{130}{10 \cdot 27} = 12 \cdot 66$ seconds.

4.7 Justification of process
The *advantages* of centreless grinding which must be considered when comparing it with more orthodox cylindrical grinding are:
1. No axial thrust is imparted to the work during the process. Long, slender parts are easily ground.
2. As deflection is minimal, heavier cuts can be taken.
3. The process is relatively simple, so fully automatic hopper feed is applicable. One operator can attend several machines.
4. Plain, short pieces can be continuously ground using the through-feed process.
5. Work centre holes are not necessary.
6. The process is relatively fast, cheap and ideal for production purposes.

4.8 Internal centreless grinding
As an appendix to this chapter it might be appropriate to briefly consider the process called internal centreless grinding. This process which is not so common as external centreless grinding requires the outside diameter first to be ground, because the internal surface is generated from the external surface.

A diagram of the process is shown at Figure 4.14.

Again the peripheral speeds of the control wheel and work are the same, and they rotate together as a pair of friction wheels. The control wheel also supports the work, in addition to the support roll and pressure roll. After the bore has been ground, the grinding wheel retracts, followed by retraction of the pressure roll. The arm under the work ejects it. (It also lowers the next component into position if automatic loading from a hopper is used.) The pressure roll then closes up to the work again. On some machines the grinding wheel spindle reciprocates because the wheel head is mounted upon slideways; on other versions the work reciprocates past a fixed position wheel spindle, the whole of the work control unit being mounted upon slideways.

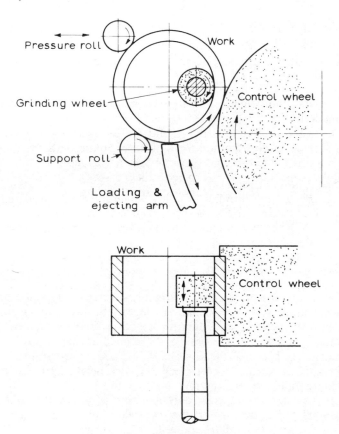

Fig 4.14 Internal centreless grinding

The process lends itself very neatly to automatic control, as the work can be loaded automatically, and if required the bore size can be gauged and controlled automatically.

The advantages are:

(*a*) high degree of concentricity between bore and outside diameter;

(*b*) high production rate, particularly where automatic loading is used. Although fixed costs of such an installation are high, variable costs are low as one operator can often 'mind' several machines.

(c) no work-holding devices required. Hence little deflection, and
    high degree of accuracy is possible even with thin walled com-
    ponents which otherwise could be troublesome;
(d) setting comparatively simple.

Learning Objectives 4
The learning objectives of this section of the TEC unit are:

4. Understands the principles and appraises the applications of centreless grinding.

4.1 Explains the principles of operation of the centreless grinding
    machine.
4.2 Describes through-feed, infeed and endfeed methods of operation.
4.3 Selects a suitable method for a given component shape.
4.4 Explains the nature and cause of lobed workpieces.
4.5 Relates machine settings to roundness errors.
4.6 Calculates through-put times for given settings.
4.7 Justifies the choice of the centreless grinding process for a parti-
    cular component.

Exercises 4

1   Explain the principles of operation of the external centreless grind-
    ing machine, making particular reference to the three major elements.
2   52 mm diameter components are to be ground by the centreless
    method on a machine having a 250 mm diameter control wheel.
    The surface speed of the work is 0·5 m/s. Calculate: (a) the work
    speed, (b) the control wheel speed, (c) work height setting. (Ans. (a)
    183·6 rev/min, (b) 38·2 rev/min, (c) 9·1 mm.)
3   With the aid of sketches describe and compare the following
    methods of operation of the centreless grinding process: (a)
    throughfeed, (b) infeed, (c) endfeed.
4   What are the main factors which determine the choice of operating
    method when components are to be ground by the centreless
    process?
5   (a) What is lobing and how is it likely to occur due to centreless
    grinding? (b) With the aid of a diagram show a satisfactory method
    of detecting lobing.
6   Plain steel spindles of 25 mm diameter × 125 mm long are to be
    centreless ground using the through-feed method. The grinding
    wheel is 150 mm wide, and the control wheel which is set at 5° to
    the work axis is 250 mm diameter. The peripheral work speed is
    0·25 m/s. Assuming no slip between the work and control wheel,

calculate (*a*) the control wheel speed in rev/s and (*b*) the cycle time/piece in seconds. (Ans. (*a*) 0·31 rev/s, (*b*) 12·7.)

7 With the aid of sketches compare the external centreless grinding process with the more orthodox external cylindrical grinding process, giving the advantages and limitations of each.

Chapter 5
Screw Thread Production

Screw threads are produced in the engineering industry in enormous numbers for a great variety of applications. Taps and dies are used in the hand processing of internal and external threads respectively, and screw threads are extensively produced upon centre lathes using single point cutting tools. This operation is covered in TEC Unit U75/001 'Workshop Processes I'. However these methods are suitable only for unit or small batch production, and other methods are used for the high production of screw threads by machining. The production machining methods described below produce accurate threads to a variety of forms, quickly and cheaply. As will be seen later in Section 5.2 the choice of method depends upon more than one factor.

5.1 Principles of production methods
The five main quantity *manufacturing methods* will be considered here, viz.,

- (a) die heads;
- (b) chasing;
- (c) rolling;
- (d) milling;
- (e) grinding.

(*a*) **Die heads.** Self-opening *die heads* or die boxes are used extensively for the high production of external threads on capstan and turret lathes and all types of automatics. The heads are containers of varying sizes which hold dies or chasers, each head being suitable for a given range of sizes. Heads are available for cutting threads from 6 mm diameter to 115 mm diameter. Chasers are available for any thread form, and using the correct type of chaser, any type of material can be screwed including some of the plastics, such as bakelite.

There are three types of die heads, the difference depending upon the types of dies used. They are those having (i) radial dies, (ii) tangential dies and (iii) circular dies. Each type is illustrated in Figure 5.1, where the diagram shows the relationship of the die blades or chasers to the work.

The dies open automatically when the required length of thread is cut. When the turret slide movement is arrested by a stop, the front part of the head continues to move forward by a small amount until the dies spring outwards, away from the work under the action of a scroll or cam. The Coventry die head having radial dies is the most commonly used general purpose die head.

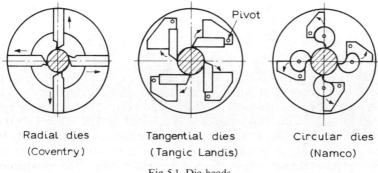

| Radial dies | Tangential dies | Circular dies |
| (Coventry) | (Tangic Landis) | (Namco) |

Fig 5.1 Die heads

The other types are used for more special applications, particularly for threading difficult materials. Provision is provided on all types for taking roughing or finishing cuts by moving a detent pin to the appropriate position. Regrinding of dies must be done in the special fixtures provided in order to maintain the correct throat and lead angles.

When used on a capstan lathe, the operator need only keep the die head up to the work by applying slight pressure to the turret slide capstan handle. The die head is self-guiding due to the action of the dies which cut at the throat only, the rest of the die thread acting as a guide nut. Hence, the die head will screw itself along the work until the dies trip open. The work spindle does not have to be reversed in order to screw the die head off again, and production rates are increased as a result. The dies can be closed by the operator after each screwing operation, by pushing a handle which partially rotates the front portion of the head. If threads of very high quality are required, the die head can be used with the chasing attachment by engaging the leader screw nut of appropriate pitch during threading.

When used on an automatic, the die head feed motion is controlled by the cam rise which is designed accordingly. After threading the dies are automatically closed by arranging for the closing handle to strike a rod on the return stroke.

Die heads are available which produce threads with a maximum pitch error varying from 0·06 to 0·30%. The cutting speeds used will be somewhat less than those for plain turning, but the following are average values (given in m/min) used for the common materials:

Brass (soft)	20
Steel (free cutting)	12
Steel (tough)	2
Cast Iron (soft)	10
Cast Iron (hard)	3

Die heads for internal threads are in fact, collapsible taps. These are similar in principle to a Coventry die head in that they have radial chasers. These withdraw, or collapse inwards when a hardened steel ring around the tap strikes the end face of the work. Hence, again there is no necessity to reverse the spindle in order to withdraw the tap, and much time is saved. They are suitable for threading holes above 25 mm diameter.

(*b*) Thread chasing. This method is similar in conception to single-point tool screw cutting on a centre lathe, but uses a *chasing tool* which is in effect several single-point tools banked together in a single tool called a chaser. The method can be carried out at speed on a capstan lathe or turret lathe equipped with an interchangeable lead screw called a leader, which is driven from the feed box. Thread chasing is usually used on production for threads which are too large in diameter for a die head. It is not confined to external threads only, but can be used for internal threads above approximately 25 mm diameter. The principle is shown in Figure 5.2.

Reference to Figure 5.2(*a*) shows a tangential type chaser being used for cutting an external thread, and Figure 5.2(*b*) shows a circular chaser being used to cut an internal thread. The system is so arranged that the chaser traverses away from the headstock whilst machining, thus preventing the possibility of the chaser being accidentally allowed to run into the chuck while traversing. For RH threads the spindle must therefore be rotated in reverse.

The chaser is advanced radially into the work for each cut by means of the cross slide screw, several passes sometimes being needed to complete the thread. This cross feed movement is independent of the quick withdrawal mechanism which operates when the leader nut is withdrawn outwards.

A standard set of interchangeable leaders and mating nuts of varying pitches are provided, which with the small range of speeds available in a capstan lathe feed gear box ensures a wide range of threads can be cut. For example, say leaders having pitches of 6 mm, 5·5 mm, 5 mm, 4·5 mm, 4 mm, 3·5 mm and 3 mm are available. With a three-speed

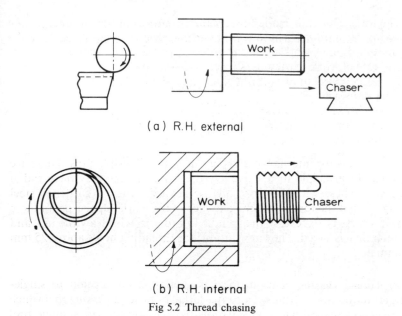

(a) R.H. external

(b) R.H. internal

Fig 5.2 Thread chasing

gear box havng ratios of 1 to 1, 1 to 2 and 1 to 4, the following pitch threads can be cut;

3 to 6 mm in steps of 0·5 mm
1·5 to 3 mm in steps of 0·25 mm
0·75 to 1·5 mm in steps of 0·125 mm

This arrangement would enable every thread pitch in the ISO metric thread standard series to be cut from 0·75 to 6 mm, with the exception of 0·8 mm pitch.

As would be expected, chasing lends itself better to non-ferrous materials rather than ferrous. Multi-start threads can be chased without any indexing of the workpiece being necessary. Taper threads can be generated by chasing, if the chasing attachment is used in conjunction with a taper turning attachment.

(c) **Thread rolling.** Threads can be *rolled* to shape by cold working in dies, and no further treatment is necessary, the thread then being complete. The process is most often applied to precision threads, but splines are also rolled to shape on production by this method. The process consists of cold flowing of the material through dies under the application of compressive forces. The material is progressively forced down to the root diameter of the thread, the excess material flowing upwards within the die space to form the upper portion of the thread.

The thread blank is made to the mean diameter of the thread, giving a saving of material. The choice of material is important, because the amount of cold working it can tolerate will depend upon its ductility and rate of work hardening. Generally speaking, metals having not less than 15% elongation and not more than 300 BHN hardness will cold work satisfactorily.

As with all cold working of metals, thread rolling produces a bonus in terms of material properties. The surface hardness, strength and fatigue resistance of the thread will all be improved. The grain flow, which follows the surface contour, will be ideal, and the surface finish will be good (approximately 0·2 μm R_a). The improvement in properties means that a cheaper material can be used for blanks in the first instance, hence reducing variable costs. The tools used are generally expensive, but die life is long and production of accurate threads can be maintained at high speeds over a long period.

Thread rolling can be accomplished using either *flat dies*, or *circular dies* contained in a die head.

Flat dies. These are used in a special production machine which can be fully automatic, requiring only the hopper filling and capable of high-speed production. The principle of operation is shown at Figure 5.3.

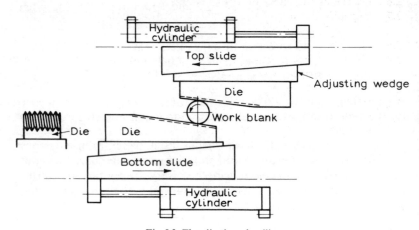

Fig 5.3 Flat die thread rolling

This diagram shows both dies moving under the action of hydraulic cylinders, although on some designs only one die traverses. The thread is formed complete after one pass of the blank between the dies. Each die is grooved with the thread profile, the thread grooves being inclined at the thread helix angle to avoid interference during rolling. The blanks can be fed automatically (or manually) from a hopper between the dies, and the die length is such that the blank rotates about four times during one pass. The dies are made from high carbon, chromium steel, and have a phenomenal life (up to 500 000 parts) due to being subjected to little friction during rolling. Rolling speeds are between 0·50 and 1·25 m/s depending upon the material properties.

Circular dies. These are contained in a die head similar to the Coventry die head, and the rolling die heads can be used upon capstan and turret lathes, or automatics. They automatically open upon completion of the thread and need not be unscrewed off the work. They are not as fast in operation as flat dies (which are almost instantaneous), but are not limited in the length of thread they can traverse. Flat dies can only roll threads which are not longer than the width of the die. A diagram of a rolling die head is shown at Figure 5.4.

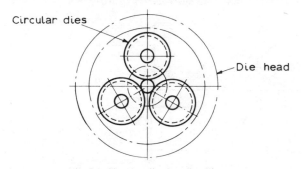

Fig 5.4 Circular die thread rolling

The dies have no cutting edges of course, but are highly polished thread rolls mounted on large, friction free spindles. The thread grooves in the rolls are not annular, but are true helical grooves. The die boxes are not suitable for large work, as the rolls must be larger in diameter than the work, say in the ratio 3 to 1. As the roll is larger in diameter than the work, the helix angle of the roll thread groove will be different than that of the work thread groove, although the pitch and thread form are the same. Hence, interference could occur on the thread flank during rolling. To avoid this, the roll thread is made multi-start in the same ratio as the roll to work diameter ratio, therefore keeping the helix angles equal. For example, if the work is 12·5 mm diameter, and the rolls are 37·5 mm diameter, then the thread rolls will have a three-start thread. In action on the capstan lathe, the die head threads itself along the work, due to the helical lead of the dies, and does not need the traversing action of a lead screw and nut. An oil lubricant is used.

High production machines incorporating independently driven

circular dies, rather than traversing flat dies, have been in use for many years. A pair of circular dies, or rollers are keyed to driven spindles which rotate in the same direction. A hydraulic cylinder controls the movement of one of the dies into, or away from, the work. The operator hand feeds the work between the dies, the threading operation then being very similar to that of plunge-cut centreless grinding. The final thread size is controlled by the infeeding die slide coming up to a pre-set stop. Higher production is obtained by use of hopper feeding.

(*d*) **Thread milling.** This is a fast production method of cutting threads, usually of too large a diameter for die heads. As the *milling cutter* is held on a stub arbor, the length of thread which can be cut is limited, although threads running up to a shoulder present no difficulty. The cutter used is called a *hob* and has annular thread grooves and form relieved teeth. A diagram of the process is shown at Figure 5.5.

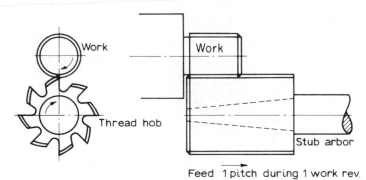

Fig 5.5 Thread milling

The hob rotates at the correct cutting speed for the operation (say 0·6 m/s), and the work slowly rotates at the correct feeding speed. As the work rotates the hob also feeds endways under the action of a master lead screw, such that as the work rotates one revolution, the hob will traverse one thread pitch. As the hob is initially fed in to full thread depth over the full length of thread, the process is complete after one revolution of the workpiece.

Although the hob has annular grooves, it is not set over at the thread helix angle, because interference is negligible at the large diameters and fine pitches used in this process. A right-hand or left-hand thread can be produced by varying the direction of hob feed and direction of rotation of work. Internal threads can be milled, although interference will then present a problem due to the longer length of engagement between hob and thread.

The machine used for this operation is relatively simple and allows a high rate of production. The work head will accommodate work-holding chucks or fixtures which can slowly rotate at the appropriate feed rate. The longitudinal feed, appropriate for the work thread pitch, is controlled by means of a lead-screw. The cutter head, which accommodates the hob, offers a range of suitable cutting speeds and has cross-wise feed to allow the cutter to be fed to the required thread depth.

Another version of the thread milling process is used for cutting threads on worms (or lead screws). This process, which makes use of a form relieved milling cutter, is less common than that using a hob described above. The operation is similar in principle to that of cutting a thread upon a centre lathe, but a revolving milling cutter is used instead of a single-point tool. Figure 5.6 illustrates the process.

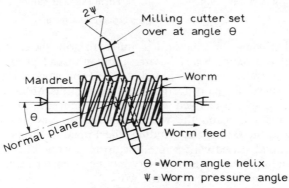

θ = Worm angle helix
Ψ = Worm pressure angle

Fig 5.6 Worm milling

The worm blank is held on a mandrel between centres (for worms with central hole) or in a collet chuck and supporting centre for solid worms. It is traversed past a rotating cutter which is carried in an adjustable head. The cutter is set over to the helix angle of the worm being cut to avoid interference. Change wheels are used to give the correct worm lead, and indexing is used for multi-start threaded worms. The cutter form must correspond to the worm tooth space in the normal plane.

Worms are often hardened and then ground, the grinding operation being identical to the milling process in principle but using a form grinding wheel in place of the form milling cutter. 0·15 to 0·25 mm is left as a grinding allowance.

Form relieved milling cutters are used, as opposed to those bearing the normal clearance angles, in order that regrinding of each tooth face may be carried out for resharpening purposes, without destroying the tooth form. Form relieving produces a spiral tooth shape behind the cutting edge, the comparison with a plain tooth shape being shown at Figure 5.7.

Normally, using this process, the milling cutter is fed in to full thread depth, the operation being completed in one full traverse of the workpiece. The form relief cutter is revolved as fast as is practical, and speeds above 40 mm/min. are used with high-speed steel cutters. Average feed rates are to the order of 100 mm/min.

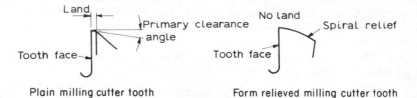

Fig 5.7 Plain and form relieved cutter teeth

(e) Thread grinding. This method is similar in principle to thread milling in that a *grinding wheel* having annular thread grooves formed around its periphery cuts the thread as wheel and work rotate. In both cases the process is one of forming and generating. Modern grinding wheel technology has enabled wheels to be produced which will maintain the thread form on the wheel, and the process therefore owes something to the skill of the grinding wheel manufacturers. A vitrified bond is generally used with a fine grit up to a value of about 600. A special grinding machine is necessary having a master lead screw and gears and the means of holding the work.

Two variations of the process are used, viz., (i) *traverse* grinding, and (ii) *plunge cut* grinding. In both cases the grinding wheel has annular thread grooves around its periphery and can be set over at the thread helix angle if necessary to avoid interference. The wheel rotates at the correct speed for the operation (say 30 m/s), while the work rotates slowly as the metal is removed.

(i) *Traverse grinding.* The principle of operation of this method is shown at Figure 5.8.

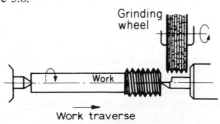

Fig 5.8 Traverse thread grinding

The wheel is positioned at full thread depth, then the work is traversed past the wheel. The worktable traverse is controlled by a master lead screw. Change gears are used to suit the thread pitch (as in screw-cutting on centre lathe). The first thread form on the wheel removes the majority of metal and is therefore subjected to the most wear. The following threads effect the finishing. A single ribbed wheel may be used for large threads or special threads.

(ii) *Plunge cut grinding.* The principle of operation of this method is shown at Figure 5.9.

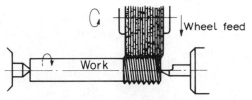

Fig 5.9 Plunge cut thread grinding

The wheel is plunged into work to full thread depth. The workpiece then makes one revolution while the work traverses one pitch. This gives uniform wheel wear but is used only for short thread lengths.

Thread grinding can be used upon soft or hardened work and cuts threads from the solid. Hence the problem of thread distortion after hardening is not present. Threads of the highest degree of pitch accuracy can be produced by grinding using machines having a master lead screw and a pitch-correction device. Clearly the accuracy of the thread profile is a function of the profile formed upon the grinding wheel face.

Wheel forming. An accurate thread profile must be produced upon the wheel face, and to achieve this different methods are used by different manufacturers. Two basic methods are (i) crushing or (ii) diamond dressing, using some suitable means to guide the diamond.

(i) *Crushing.* This makes use of a crushing roller of hardened steel which has the required thread form accurately produced around its periphery and is ground and lapped. The roller is fed into the wheel face under pressure using a great volume of lubricant, the thread profile being crushed into the wheel face, while the wheel rotates slowly.

The principle of operation of this method is illustrated at Figure 5.10.

The crushing unit may be mounted on the wheel head above the wheel as shown, or on some machines the attachment is mounted upon the table. Care has to be taken otherwise excessive loads may be

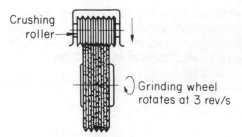

Fig 5.10 Grinding wheel crushing

placed upon the wheel bearings. Crushing is a forming process as opposed to dressing the thread profile upon the wheel which is a generating process.

(ii) *Diamond dressing.* Different systems may be used for generating the thread form upon the wheel face by dressing. For multi-ribbed wheels the patented device used upon 'Matrix' machines provides an interesting example. This is shown in simple diagrammatic form at Figure 5.11.

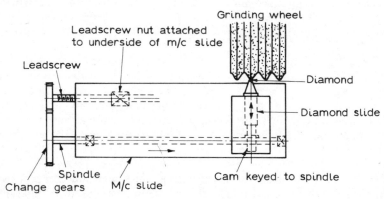

Fig 5.11 Diamond dressing attachment

As the lead screw is driven from the machine, the cam rotates upon the spindle which is coupled to the lead screw through change wheels. The cam imparts the longitudinal movement to the diamond. A suitable combination of cam lift and change wheels will give the desired thread profile upon the wheel face, such that as the machine slide traverses the diamond past the wheel, the lead screw rotation will cause the diamond (via the cam) to trace out the thread profile.

For larger thread profiles such as required upon a single-ribbed wheel, a pantograph of the type shown at Figure 5.12 may be used.

The pantograph shown is a mechanical tracing device in which the

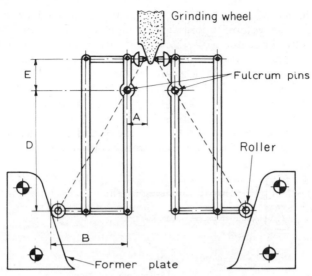

Fig 5.12 Pantograph wheel dressing device

follower roller centre, fulcrum axis and diamond tracer tip all lie on a straight line (shown dotted in Figure 5.12).

$$\text{The pantograph ratio} = \frac{D}{E} = \frac{B}{A}$$

This ratio varies usually from 5:1 to 10:1, hence a large accurate profile on the former plate can be reproduced on the grinding wheel on a reduced scale consistent with the pantograph ratio. On the device shown the roller is manually followed around the form on the plate, and the diamond follower automatically traces out, and dresses, a similar reduced form upon the wheel. The great advantage of the system shown is that the wheel can be lowered and re-dressed at any time without disturbing the pantograph unit. Hence wheel wear does not effect the accuracy of the finished thread profile. Commercial pantograph wheel-forming devices are available which are based upon the same principle but are of a much more sophisticated design. They will accurately reproduce virtually any required form from a template on to a grinding wheel.

5.2 Choice of method
Consideration will be given here to the *advantages* of each of the methods described in Section 5.1, with a component shown as a typical example of threads produced by each method respectively.

(*a*) **Die heads.** Much general purpose externally turned work is produced in large volume upon capstan or turret lathes, or automatic machines from bar stock; often in free cutting metal. Where such work bears a thread upon the end, a die head is the ideal threading medium. It is relatively easy to set-up, maintain and sharpen the blades, and is quick and easy in use. It therefore lends itself to the quick production of standard threads upon such parts as bolts, screws, spindles, etc.

Figure 5.13 shows a tee-bolt which is required in large quantities and is produced upon a capstan lathe. The thread would readily be cut using a die head, and the required tolerance could easily be maintained.

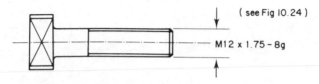

(see Fig 10.24)

M12 x 1.75 – 8g

Material:– free-cutting mild steel

Fig 5.13 Tee-bolt

(*b*) **Thread chasing.** This method is an ideal alternative to that of die heads, for quantity production of threads which are not easily produced by die heads. It has particular advantages for large diameter work, both internal and external, and is particularly suited to non-ferrous materials. It is not prone to clogging up by metal cutting chips, as might a die head under certain circumstances. Although thread chasing can be carried out at remarkable speed, more than one pass is often necessary to complete the thread to full depth.

The knurled nut shown at Figure 5.14 is required in large quantities, and the internal thread would lend itself to production by a thread chasing operation.

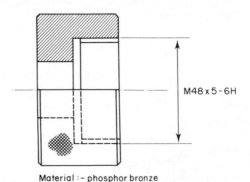

M48 x 5 - 6H

Material :– phosphor bronze

Fig 5.14 Knurled nut

(*c*) **Thread rolling.** This process lends itself to the high production of external threads particularly those requiring high fatigue resistance. A saving in material costs is possible when compared to other methods, and the degree of precision and surface finish obtainable is excellent. Indeed, micrometer screw threads have been produced satisfactorily by rolling. The work material used must of course lend itself to deformation by cold working. Because of the high pressures needed for flow forming, the process is generally restricted to fine pitch threads and relatively small diameters up to about 20 mm.

The stud shown at Figure 5.15 could have the threads required at both ends produced at high speed and efficiency upon blanks fed into a thread rolling machine. Such a component requires mechanically strong threads, and the material cost savings could be considerable when large quantities are required.

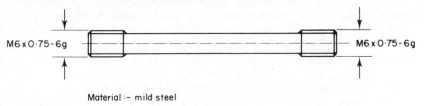

Material :— mild steel

Fig 5.15 Stud

(*d*) **Thread milling.** The version of this process which uses a form relieved milling cutter is possibly the only practical way of producing large, special thread forms upon components such as worms or lead screws to the required degree of accuracy.

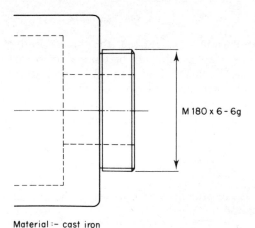

Material :— cast iron

Fig 5.16 Threaded component

The more common version of this process, which makes use of a cutter called a hob, is a useful and efficient alternative method to those already considered. It can be used for cutting internal or external threads which are too large a diameter, or too coarse a pitch for other methods. It produces smooth and accurate threads, and is often the only practical way of producing threads which butt up to a shoulder. The process can cope with the harder, tougher ferrous materials which may be difficult for other processes.

Figure 5.16 shows a thread produced upon the end of a component which could most efficiently be milled on a production machine.

(*e*) **Thread grinding.** This is a versatile and highly accurate method of producing external threads, with a pitch error of not more than 0·01% being possible. It can produce fine or coarse pitch threads from the solid, or can be used for finishing pre-formed threads. Components hardened by heat-treatment can be thread ground. Many screwing taps are finish ground after the threads have been pre-formed by rolling. This combination of processes is said to give a larger life to the product. The process is versatile because it could be used for producing medium quality threads on studs in large quantities, for example, or for producing the highest quality threads on, say thread gauges.

The hardened screwing tap shown at Figure 5.17 would be a typical component produced in quantity by grinding from the solid.

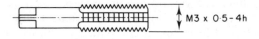

Material :- hardened tool steel

Fig 5.17 Tap

In summary it should be said that the thread production methods described in this chapter can only be compared to each other in the most general way. Each has its own range of applications to which it is most suited, but much threaded work could be produced equally well by more than one method. In such cases the criteria might then become production cost, expediency, availability of equipment, etc.

Learning Objectives 5
The learning objectives of this section of the TEC unit are:

5. Understands the principles and appraises the applications of screw thread production methods.

5.1 Describes and explains the principles of the following methods of screw thread production: Thread chasing, rolling, milling, grinding.

5.2 Selects and justifies a particular method, from those listed in 5.1, for a given type or class of screw threaded component.

Exercises 5

1 With the aid of sketches show how a 40 mm diameter, left-hand, internal thread may be chased upon a capstan lathe. Show the direction of rotation of the spindle, the position and feed direction of the chasing tool. Sketch the chaser.

2 Describe and compare the following three methods of producing external threads on components in quantity: (*a*) Coventry die head, (*b*) chasing, (*c*) rolling. State the advantages and disadvantages of each method.

3 Using diagrams to illustrate your answer, explain the principle of operation of thread milling for (i) fine pitch threads, and (ii) coarse pitch threads such as worms.

4 Using diagrams to illustrate your answer, explain the principle of operation of thread grinding using: (*a*) the traverse grinding method, (*b*) the plunge-cut grinding method.

5 Two main methods are used for imparting a thread profile upon the abrasive wheel of a thread grinding machine, viz., diamond dressing and crushing. Describe and compare each of these methods.

6 Name and explain the main factors which must be taken into consideration when selecting a particular screw thread production method.

Chapter 6
Broaching

Broaching is an interesting metal-cutting process which is used for quantity production, and can produce accurate plane or formed shapes on internal or external surfaces. Metal is removed by the successive action of a number of cutting teeth incorporated in a tool called a broach. One pass of the broach completes the operation. Machines may be of horizontal or vertical configuration, and the process is simple in conception. Although broaching is used extensively to produce simple shapes, it is often the only practical method of machining complicated internal or external profiles in production quantities.

Internal broaching is used to produce round, square, hexagonal and many other shapes of holes. Internal keyways, splines and gears for example are easily produced by broaching.

External, or surface broaching, can also produce plane or complex shapes and is increasingly used for production purposes as an alternative to milling. At one time most broaching was internal, but surface broaching is now used just as frequently.

6.1 Broaching machine
The function of a *broaching machine* is to provide the means for locating and holding both the work and the broach, and to give the necessary forward and reverse broach movement (stroke), at the required cutting speed. For convenience, broaching machines can be classified according to whether they are of horizontal or vertical configuration, thus:

1. Horizontal machines, may be
   (a) mechanically operated by screw and nut;
   (b) hydraulically operated;
   (c) electro-mechanically operated using a rack and pinion drive controlled by a Ward–Leonard electrical system (as used on planing machines);
   (d) push or pull type, the latter being most common;
   (e) used for surface or internal broaching, the latter being most common.

2. Vertical machines, may be
   (*a*) hydraulically operated, this almost always being the case;
   (*b*) push-up, push-down, pull-up or pull-down type;
   (*c*) used for surface or internal broaching, the former being most common.

The speed of operation of machines is increased if required by the use of indexing worktables, automatic fixtures, etc. Duplex machines are available, these having a double ram arrangement such that one is engaged on the cutting stroke as the other ram is returning. Hence, one work station will be unloaded and re-loaded whilst the other station is having work machined, and output is approximately doubled. Special broaching machines have been designed for certain operations, such as gun-bore rifling, and drum or turret broaching machines are in use, but these classes of machines will not be considered here.

The most common mode of operation for modern broaching machines is hydraulic, and the maximum ram force available is to the order of 500 KN, with a maximum cutting speed of about 10 m/min and return speed of 60 m/min. Electro-mechanically operated machines, (mentioned under 1(*c*) above), developed exclusively for surface broaching have capacities well in excess of hydraulic machines.

In order to illustrate the principle of operation of broaching machines, three types will be described, viz;

mechanical horizontal machine (ref. 1(*a*) above);
hydraulic horizontal machine (ref. 1(*b*) above);
hydraulic vertical machine (ref. 2(*a*) above).

Mechanical horizontal machine. A diagram of such a machine being used to carry out an internal pull-broaching operation is shown at Figure 6.1.

This outlines the essential elements of what is basically a simple machine tool. A rotating nut is housed at the drive end of the machine, and the speed of rotation of the nut (and hence the broach cutting speed) is controlled through a gear-box. A fast return, non-cutting stroke speed can also be selected. The rotating nut draws the screw through it, and in turn the broach (which is attached to the screw ram by means of an adaptor or puller) will be drawn through the work. The diagram shows adjustable trip-dogs which can be set to automatically terminate the traverse as required. The technique of internal broaching will be covered in more detail in Section 6.2, when Figure 6.1 will again be referred to.

Hydraulic horizontal machine. A sectioned diagram of a typical machine of this type is shown at Figure 6.2.

The machine shown has an infinitely variable range of cutting and

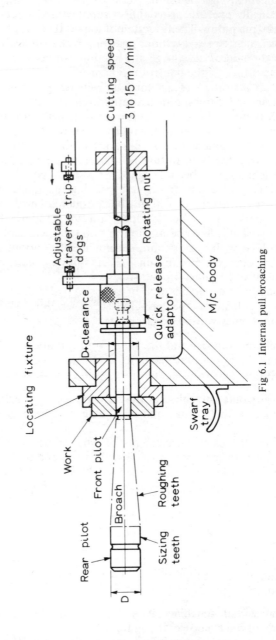

Fig 6.1 Internal pull broaching

return speeds, with the maximum return speed being much greater than the cutting (broaching) speed. Oil is supplied under pressure by a variable-discharge pump. The direction of the oil flow can be reversed as required on such a pump without the need to change its direction of rotation, i.e. the pump's discharge and suction can be reversed manually or by trip stops on the control mechanism.

Figure 6.2 shows the machine starting on its return stroke, with the pump delivering oil to the quick-return cylinder in the bore of the piston rod thus moving the rod from right to left. During this return stroke the oil in the L.H. end of the main cylinder is expelled, part of it returning to pump suction (as shown) and the rest being drawn by the moving piston into the R.H. end of the main cylinder. Additional oil is

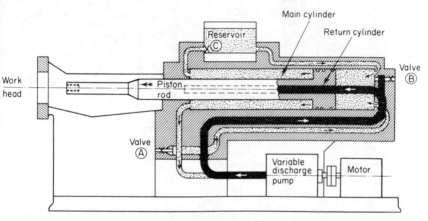

Fig 6.2 Horizontal hydraulic broaching machine

required by the R.H. cylinder end and this is drawn from the reservoir. Throughout this stroke, valves *B* and *C* are closed. At the end of the stroke, with the broach and work in position ready for cutting, the oil flow is reversed.

Oil now flows from the discharge end of the pump to the L.H. end of the main cylinder, the piston rod moving from left to right on the cutting stroke. Valve *A* is automatically actuated and closed, shutting off the by-pass to the R.H. end of the main cylinder, and the entire pump discharge is therefore being delivered to the L.H. cylinder end. Part of the pump suction is being taken from the return cylinder, and part from the R.H. end of the main cylinder with valve *B* open.

Hydraulic vertical machine. Figure 6.3 illustrates the hydraulic system for a machine of this type.

The machine shown has a vertical slide to which the broach is at-

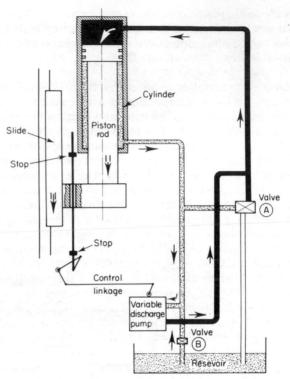

Fig 6.3 Vertical hydraulic broaching machine

tached, the slide having an infinitely variable speed range for both the cutting (downward) stroke and return stroke. Note that because of the large diameter piston-rod, a considerable volume differential exists between the upper and lower ends of the cylinder.

Figure 6.3 shows the machine starting on its downward stroke as oil is delivered from the discharge end of the pump to the upper end of the cylinder, hence forcing the piston downward. Oil drains from the lower end of the cylinder to pump suction. Because of the volume difference, the additional oil required at pump suction is drawn from the reservoir as shown, with valve *B* open. At the end of the downward stroke, the slide engages the lower adjustable stop thus moving the control linkage which reverses the flow of oil from the pump.

Oil now flows from the discharge end of the pump to the lower end of the cylinder, the piston (and hence broach) returning rapidly because of the smaller volume. The volume of oil expelled from the upper end of the cylinder is greater than required by pump suction. However, the pressure build-up created in the delivery line to the pump actuates valve *A*, which opens and by-passes the excess oil to the reservoir, valve *B* of course being closed.

6.2 Broaching techniques

The *technique of broaching* as a means of producing accurate profiles of fine finish can be applied to *internal* and *external* surfaces. Each will be considered in turn.

Internal broaching. Metal is removed by the successive action of a number of cutting teeth incorporated in a tool called a broach. The workpiece is located and held (not usually clamped) in the machine, and the broach is either pushed or pulled through a previously rough machined hole in the workpiece. An operator is required to thread the broach through the work and couple it to the machine ram (for pull broaching), but the process is still a fast way of producing accurate holes of any shape. Figure 6.1 shows a pull broaching operation in which an internal broach is being pulled through a workpiece firmly held in the locating fixture under the influence of the cutting force. The roughing teeth successively remove the material before the sizing teeth finish the hole. The front pilot (clearance fit in the starting hole) and rear pilot (clearance fit in the finished hole) support the broach at the start and finish of the operation. One stroke of the machine completes the operation. In the case of a non-round hole (say square for example) the broach would change the initial round hole to a square hole in the one operation. If a situation arises where the broach would be unacceptably long in order to accommodate the required number of teeth, then more than one broach must be used. Vertical push-down broaching is a simple operation in which the broach is located in the workpiece, and then pushed through by the machine ram. Long, slender broaches do not lend themselves to this operation.

In pull broaching as illustrated at Figure 6.1, a quick and convenient way must be found of attaching the broach to the machine ram. The quick-release adaptor, or puller shown is commonly used for round broaches. Figure 6.4 shows a sectional view of this type of puller in the closed, or locked position.

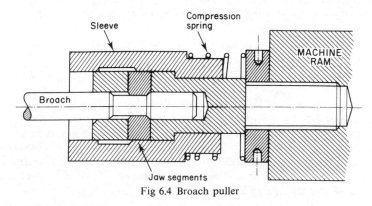

Fig 6.4 Broach puller

In the locked position shown, the hardened jaw segments are securely closed in on the broach neck as the machine ram pulls the broach through the work. To release the broach at the end of the return stroke, the outer sleeve is moved to the right against the spring, either by hand or machine stop. The broach can then be pulled clear as the jaw segments retract into the sleeve inner clearance groove. The broach can be reloaded into the puller in similar fashion after threading through the work.

An alternative and more simple pulling system makes use of a taper cotter, as shown at Figure 6.5.

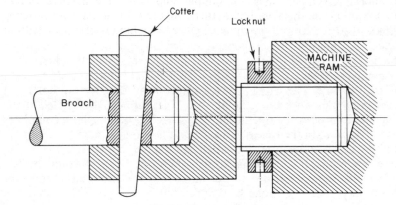

Fig 6.5 Broach puller

This system is used extensively for larger broaches which must have a shank of sufficient diameter to accommodate a slot to suit the cotter.

Accurate internal keyways required in quantity can easily be broached, and this is a common form of internal broaching. The set-up is shown at Figure 6.6.

The flat sided keyway broach is supported by a packing piece, or

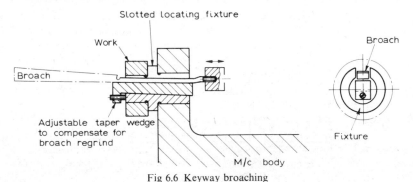

Fig 6.6 Keyway broaching

taper wedge which is located in the bottom of a guide slot in the fixture. The keyway is cut by teeth on the top face only of the broach. This type of broach usually has a screwed shank which is attached to a corresponding threaded puller.

External broaching. The broaching of external work surfaces is known as surface broaching, and the process is similar to that of internal broaching, but the broaching force will not hold the work in position on the fixture but will push it away. Therefore, surface broaching fixtures are more elaborate requiring clamping arrangements. The process is a direct alternative to milling, and hence the fixtures will be similar in principle to milling fixtures. Figure 6.7 is a diagram showing the set-up for surface broaching. Surface broaching is often carried out

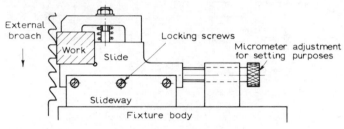

Fig 6.7 Surface broaching

on relatively large components, and the rise/tooth may be as high as 0·25 mm or more, and broaches may be stellite or tungsten carbide tipped.

The following points, in addition to those already covered under internal broaching, should be considered for surface broaching.

(a) Fixtures should be rigid, quick acting with fine adjustment for positioning work relative to broach cutting edges built in to the fixture. Automatic operation for long runs should be considered, or a duplex (double slide) machine used with two fixtures. In the latter case one fixture will be loaded and unloaded as the other is in use. To avoid idle time the unloading and loading time should be less than the cutting time.

(b) The work and fixture should have no obstructions in the plane of the surface to be broached.

(c) The work and clamping must be strong enough to withstand the eccentric forces placed upon them during cutting.

(d) In theory any shape of external surface can be broached, but in some cases two operations may be required in order to simplify the broach design. Such a case is shown at Figure 6.8.

It will be noticed that the broach is made up in segments, inserted and locked into position in a holder. This facilitates manufacture and re-grinding, and also means that a broken broach need not be scrapped. A progressive cutting action can be given to the broach teeth by inclining the cutting edge at an angle to the direction of travel. This is comparable to using a helical slab milling cutter as opposed to a straight fluted milling cutter.

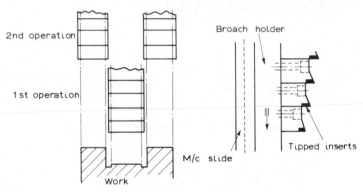

Fig 6.8 External broaches

6.3 Justification of process

In general broaching is a process which can produce plain or complex profiles to a high degree of accuracy and finish at high production rates. In certain cases complicated internal profiles required in large quantities can only sensibly be produced by broaching. The following *advantages* are claimed for broaching:

(*a*) Fast and simple process;
(*b*) High degree of accuracy and finish possible;
(*c*) Virtually any shape of profile can be machined;
(*d*) As cutting force per tooth is low, tooth wear is low. A broach can generally produce more parts between tool re-grinds than any other type of cutting tool;
(*e*) Cutting temperatures are low as the cut per tooth is small;
(*f*) Lends itself to semi- or fully automatic operation.

The following limitations apply to broaching:

(i) A relatively expensive process suitable only for high quantity production;
(ii) The manufacture of broaches (particularly internal types) is a specialised and expensive operation;

(iii) No features must be present which cause obstruction in the plane of the surface to be broached;
(iv) The work must be strong enough to withstand the cutting forces;
 (v) Large amounts of metal cannot be removed by broaching.

For internal work broaching may be considered as an alternative process to reaming, milling, generating processes (such as gear cutting), moulding, etc., and for external work as an alternative to milling, shaping, etc. When comparing one production process, such as broaching, with another, such as milling, the main criterion to consider is usually the manufacturing cost. A useful tool for cost analysis is the break-even chart, and an example is shown at Figure 6.9 where broaching and milling are under consideration as suitable alternatives for the production of a particular component.

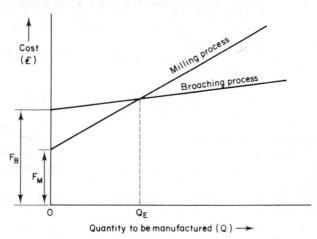

Fig 6.9 Break-even cost chart

Break-even charts have the fixed and variable costs for each process represented to scale. Fixed costs are independent of the quantity of the product to be manufactured, and include the cost of tooling, fixtures, setting-up, etc. In Figure 6.9 the broaching process fixed costs are F_B, being higher than the milling process fixed costs F_M. Variable costs vary proportionally to the quantity of products made, i.e., the slope of the total cost line increases as the variable costs increase, where total cost is the sum of the fixed and variable costs. Variable costs include the costs of labour and materials. In Figure 6.9 the broaching variable costs are less than the milling variable costs, the slope of the broaching total cost line being the lesser.

If a quantity greater than the break-even quantity Q_E is required, then the product could more economically be produced by the broaching process.

6.4 Design factors

Figure 6.10 illustrates a typical internal pull-type broach for machining a hole to a final shape from a suitably sized starting hole, which is usually round.

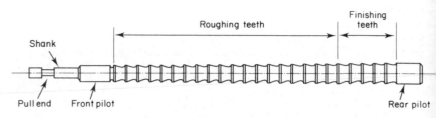

Fig 6.10 Internal broach

The main factors affecting the *design* of such a broach are considered below, although in practice a design study would involve other details.

Broach material. High-speed steel is the most suitable broach material, a typical composition being 0·7% carbon, 18% tungsten, 4% chromium, 1% vanadium. The HSS is hardened before the teeth are finish ground.

Tooth shape. This is shown for roughing and finishing teeth at Figure 6.11(*a*) and (*b*).

Burnishing teeth, which are non-cutting teeth, are occasionally included in an attempt to improve the surface finish.

Tooth Pitch (*P*). As a general rule the pitch should be such that at least two (preferably three) teeth are cutting at any one time. The following empirical formulae will be found satisfactory:

$$\text{Pitch } P = 1 \cdot 77 \sqrt{\text{length of hole}}$$

Very thin workpieces will be better clamped together and broached as a solid piece.

Rise/tooth (*t*). Depends upon:
(*a*) Shape of hole.
(*b*) Type of material being broached.
(*c*) Force available at the machine.
(*d*) Size of hole.
'*t*' is generally quite small, of the order of 0·025 mm to 0·160 mm.

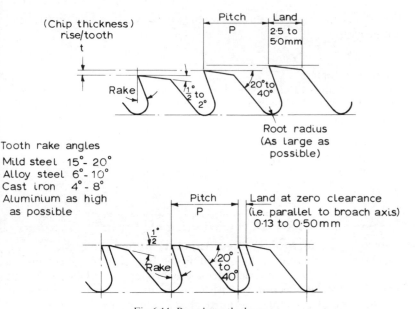

Fig 6.11 Broach tooth shape

Length of cutting portion of broach (L).

$$L = (P \times \frac{T}{t}) + (P \times S) = P\,(\frac{T}{t} + S)$$

where T = Metal to be removed by roughing teeth.

S = Number of finishing teeth (4 to 6).

Note that $\dfrac{T}{t}$ represents the number of roughing teeth and should be a whole number.

Load on Broach (F). Generally F = (Area of metal removed by the teeth in contact with the work) × (Force to remove 1 mm² of metal at a given rise/tooth). The latter parameter is usually designated by K and is tabulated on p. 78. It is not the shear strength of the material being broached.

For round holes F = hole circumference × N × t × K
$\qquad\qquad\qquad = \pi dNtK$ where d = finished hole diameter
$\qquad\qquad\qquad\qquad\qquad N$ = maximum number of teeth
$\qquad\qquad\qquad\qquad\qquad\qquad$ cutting at once.

For square holes F = hole perimeter × N × t × K
$\qquad\qquad\qquad = 4HNtK$ where H = finished length of
$\qquad\qquad\qquad\qquad\qquad\qquad\qquad$ one side of square.

Rise/tooth (mm)	K value (Newtons)		
	Cast iron	Low-carbon steel	Alloy steel
0·025	3500	3850	5000
0·040	2750	3150	4150
0·050	2350	2650	3600
0·060	2150	2350	3400
0·075	1950	2150	3150
0·100	1650	1950	2850
0·120	1600	1800	2800
0·140	1500	1750	2500
0·150	1500	1700	2450
0·180	1500	1650	2350

In practice, after calculating 'F' in order to find the capacity of the machine capable of pushing or pulling the broach through the hole, it would be necessary to calculate if the broach is sufficiently strong across its weakest section to withstand 'F' in tension or compression.

The general principles outlined above also apply to the design of surface broaches.

6.5 Broach design

As a simple *design study* of a broach, consider the following example in which a drilled hole is to be completed to its final diameter and finish by pull broaching.

Example 6.1

The bore of an alloy steel component is to be finish broached to $31·75^{+0·01}$ mm diameter, the bore prior to broaching being $31·24^{+0·5}$ mm diameter. Calculate (*a*) pitch of teeth, (*b*) length of cutting portion and (*c*) force to pull broach through work, if length of hole = 25 mm, $t = 0·025$ mm, $K = 5000$ N and $S = 5$. Sketch the broach.

Solution
(*a*) $P = 1·77 \sqrt{\text{hole length}} = 1·77 \sqrt{25} = 8·85$ mm say 9 mm
Check. This gives a minimum of two teeth cutting in the worst condition, therefore 9 mm is satisfactory.

(*b*) $L = P(\dfrac{T}{t} + S)$ where $T = \dfrac{31·76 - 31·24}{2} = \dfrac{0·52}{2}$

$$= 0·26 \text{ mm}$$

$$\therefore \qquad L = 9\,\frac{0.26}{0.025} + 5 = 9\ (say\ 11 + 5)$$

$$= 9 \times 16 \qquad = 144\ mm$$

The first cutting tooth will be 31·24 mm diameter, and the second will be 31·26 mm diameter. The rest of the cutting teeth will rise in steps of 0·05 mm on diameter, and the finishing teeth will be 31·76 mm diameter.

In practice up to 0·05 mm on diameter might be allowed for hole shrinkage after broaching.

(c) $\qquad F = \pi d N t K$

$\qquad\qquad = 3.142 \times 31.76 \times 3 \times 0.025 \times 5000$

$\qquad\qquad = 37.40\ KN$

$\qquad\qquad$ Choose a 40 KN capacity machine.

A sketch of the broach is shown at Figure 6.12.

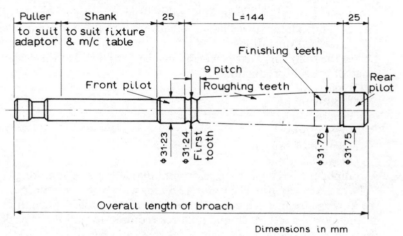

Dimensions in mm

Fig 6.12 Round internal broach

Learning Objectives 6
The learning objectives of this section of the TEC unit are:

6. Understands the principles and appraises the applications of broaching.

6.1 Explains the operating principle of the broaching machine.

6.2 Describes the techniques of internal and external broaching.

6.3 Justifies the use of the broaching process with respect to other methods of manufacture.

6.4 Identifies factors affecting broach design.

6.5 Calculates tooth spacing, number of teeth, maximum load and designs a simple broach from given data.

Exercises 6

1 With the aid of sketches describe the operating principle of a horizontal broaching machine having either a hydraulic, or mechanical mode of operation.

2 Sketch and describe the principle of operation of a hydraulic vertical broaching machine.

3 Produce a diagram of a round pull broach, and a typical puller which might be used with it.

4 Sketch a typical pull keyway broach, puller and fixture, suitable for a keyway broaching operation.

5 With the aid of a suitable diagram, describe the external broaching operation.

6 Compare, stating the advantages and limitations, the processes of internal and external broaching respectively with other comparable metal removal operations.

7 Define the following factors which must be considered at the design stage of broach production:
    (*a*) material,
    (*b*) tooth shape,
    (*c*) tooth pitch,
    (*d*) rise/tooth,
    (*e*) length of broach,
    (*f*) load on broach.

8 A component having a bore of 50 mm diameter × 75 mm long is to have a keyway broached in it 12·5 mm wide × 6·4 mm deep on the bore centre-line. Calculate (*a*) broach tooth pitch, (*b*) length of cutting portion of broach and (*c*) force required to pull the broach through the work, if rise/tooth = 0·1 mm, number of sizing teeth = 6, and K = 3000 N. Sketch the broach which is to be used upon a horizontal machine, and also sketch a quick release adaptor suitable for coupling the broach to the machine ram. (Ans. (*a*) 16 mm, (*b*) 1·12 m, (*c*) 18·75 KN.)

Chapter 7
Gear Cutting

Spur gears are used to connect two parallel shafts, the smaller gear in the train being called the pinion, and the larger the wheel. Consideration of gear cutting methods in this chapter will be confined to spur gears. The tooth shape most commonly used for gears is based upon the involute curve. Figure 7.1 shows that the end of a tightly wrapped cord unwound from a cylinder will describe an involute curve.

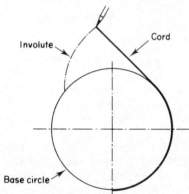

Fig 7.1 Construction of involute curve

If the cylinder is looked upon as a base circle, then the involute curve will vary as the base circle size varies. As the base circle diameter increases, so the involute curve gets straighter until it becomes a straight line generated from a circle of infinite radius. Hence, one has the basis of a rack having straight-sided teeth. Figure 7.2 shows the elements of a spur-gear involute tooth form, and Figure 7.3 a rack involute tooth form.

Figure 7.2 is a detail of part of two meshing gears A and B, their reference circles touching at point P. This is called the pitch point. If two discs were made to the diameters of the reference circles, one could

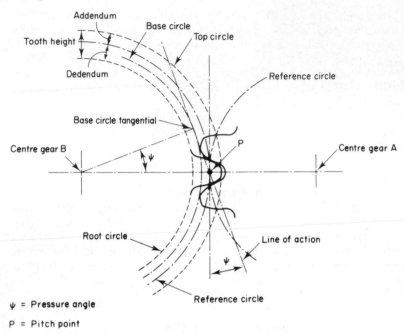

ψ = Pressure angle

P = Pitch point

Fig 7.2 Involute gear tooth form

be made to drive the other as friction wheels. However, in order to generate the required power, interlocking teeth are required, and the teeth in modern gears have flanks of involute form. This provides uniform motion between the gears.

The involute curves of the teeth flanks are generated from the base circle, between the limits of the root circle and top circle. A suitable root radius completes a tooth shape. Adjacent tooth flanks make point contact on the involute curves as the gears rotate, contact being made on the line of action. This is a straight line passing through pitch point

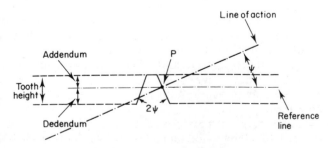

Fig 7.3 Involute rack tooth form

P, and is inclined at angle ψ which is known as the pressure angle. This angle affects the involute shape. However, ψ is usually 20° on modern gears which seems to give optimum strength and running conditions.

Figure 7.3 shows a rack which is a gear of infinite diameter so that the reference circle becomes a straight line. The tooth flank is straight and inclined at angle ψ. Hence, the tooth included angle is equal to twice the pressure angle.

Teeth are spaced out along the reference circle or reference line at equal increments known as the pitch. The metric pitch is called the module (*m*), this being the length in millimetres that each tooth would occupy if the teeth in the gear were spaced along the diameter of the reference circle (*D*).

Example 7.1

Find the module of a gear of 150 mm reference diameter having 50 teeth.

Solution

$$m = \frac{D}{T} \qquad \text{where } m = \text{Module.}$$

$$D = \text{Diameter of reference circle.}$$

$$T = \text{Number of teeth in gear.}$$

$$\therefore \qquad m = \frac{150}{50} = 3 \text{ mm}$$

Tooth height is equal to the tooth addendum (Add.) and dedendum (Ded.) where the addendum is the radial distance from the reference circle to the top circle. The dedendum is equal to the addendum plus suitable tooth clearance, this being 0·25 of the module. The tooth addendum is equal to the module, hence:

$$\text{Add.} = m$$
$$\text{Ded.} = 1\text{·}25 \, m$$

$$\text{Tooth height} = \text{Add.} + \text{Ded.} = 2\text{·}25 \, m$$

Example 7.2

Find the addendum, dedendum and working height of a gear tooth of 3 mm module.

Solution

$$\text{Add.} = m = 3 \text{ mm}$$
$$\text{Ded.} = 1 \cdot 25\, m = 3 \cdot 75 \text{ mm}$$
$$\text{Tooth height} = \text{Add.} + \text{Ded.} = 3 + 3 \cdot 75 = 6 \cdot 75 \text{ mm.}$$

At this stage it should be noted that two gears of the same module and pressure angle will mesh with each other, and each gear will also mesh with an involute rack of the same module and pressure angle. The fact underlies the principle of gear generation which is the method used to produce accurate spur gears in quantity, and will be considered in the subsequent sections of this chapter. However, many gears are still produced by the forming method which we will briefly consider to conclude this introduction.

Gear Forming. This method is confined to the production of single, or small numbers of gears. Gear forming is carried out on a horizontal milling machine using a cutter bearing the profile of the tooth space, this cutter being fed into the blank to the correct cutting depth to give the required tooth height. The operation is shown at Figure 7.4.

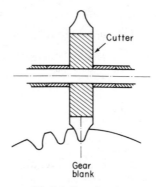

Fig 7.4 Gear forming

The gear blank is secured to a dividing head so that the tooth pitch is accurately maintained. Each tooth space is separately gashed out so that the process is relatively slow. For spur gears the tooth spaces (and hence the tooth flanks) are parallel to the axis of the gear.

The tooth space changes as the module, pressure angle and reference circle diameter changes, so theoretically a large number of different shaped cutters would be required to cover a reasonable modular range of gears having varying numbers of teeth. In practice a compromise is effected by using a set of eight cutters for each module, so that the tooth space shape is only approximately correct. Gears produced by forming are satisfactory under moderate running conditions.

7.1 Spur gear manufacture

Generating is the method used to *produce* accurate gears in *large quantities* at speed. In generating processes the cutting tool shape has no direct effect upon the finished workpiece profile. The form of the profile is produced by the cutting edge of the tool being constrained to move through the required path to achieve the generated shape. With forming, each tooth is separately and individually formed as each tooth space is gashed in the blank; with generating the teeth are machined in one continuous process.

The three basic processes used for gear generation are planing, shaping and hobbing. In each case one starts with a plain disc as a blank which is usually mounted upon a mandrel or arbor. In this section the operating principles of the gear cutting machines will be described.

Planing machine. Gears can be planed using the Sunderland process or the Maag process, which are identical in principle but different in the machine configuration and detail. Figure 7.5 shows the essential requirements of a Sunderland type machine, which utilises an involute-rack type cutter.

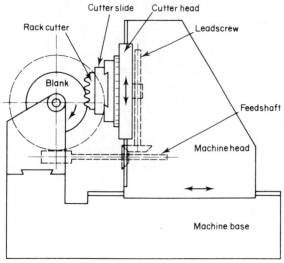

Fig 7.5 Gear planing machine

The blank is shown mounted upon an arbor, and the rack cutter held in a tool-box attached to the cutter slide. The cutter slide reciprocates parallel to the axis of the blank to effect the cutting. The cutter slide can be swivelled and set at an angle on the cutter head, in order to cut helical gears. The cutter head can slide up and down slideways on the machine head, which itself can slide along slideways on the machine base. This latter facility enables the rack cutter to be fed in to the required depth of cut.

The sliding motion of the cutter head is controlled by a master lead screw, which in turn is connected through gearing and a worm and worm-wheel to the blank arbor. Change wheels, not shown in the simplified diagram at Figure 7.5, are connected into the drive between the cutter head and the blank arbor. Hence, the appropriate wheels can be fitted to give the desired relationship between the blank rotation and the cutter straight-line movement. It should be noted here that it is possible to disconnect this gearing when required during the cutter 'stepping-back' operation, which is explained in Section 7.2.

Shaping machine. Gears can be shaped using the Fellows process which makes use of a pinion-type cutter. The essential elements of a Fellows-type machine are shown at Figure 7.6.

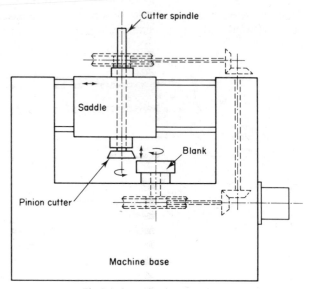

Fig 7.6 Gear shaping machine

This figure shows the blank mounted upon a vertical arbor, and a pinion cutter fixed to the cutter spindle which is carried by the saddle. A guide controls the action of the cutter spindle which is reciprocated by a crank and operating arms (not shown). A helical guide must be fitted for the production of helical gears. The saddle can be fed along slideways as required.

The cutter spindle and the work spindle are connected to each other through gearing, as shown. Not shown, however, are change wheels which are connected into this drive. Hence, the appropriate wheels can be fitted to give the desired gear ratio (velocity ratio) between the blank and pinion cutter. This obviously depends upon the number of teeth on the cutter, and the number of teeth required upon the blank.

Hobbing machine. Gears can be hobbed upon a machine utilising a cutter called a hob. This is of worm-like configuration and is used in the same way as a milling cutter. Figure 7.7 outlines the essential elements of a hobbing machine.

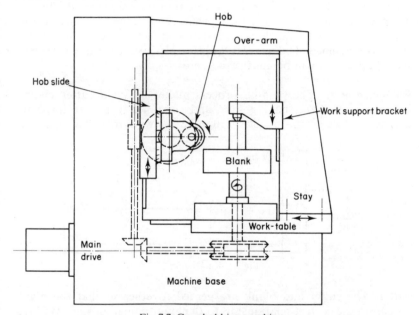

Fig 7.7 Gear hobbing machine

The machine shown at Figure 7.7 is of vertical configuration, although horizontal machines are also used. The blank is mounted upon a vertical arbor which is carried on the worktable, the arbor being supported at the top by a bracket. The bracket is mounted upon slide-ways on the vertical stay, which in turn is fastened to the worktable. An over-arm gives added stiffness to the stay, ensuring complete rigidity during cutting. The worktable is also mounted upon slide-ways in order to accommodate varying size blanks. The hob is mounted upon a spindle which is carried in the hob slide which can be traversed vertically in either direction. The head carrying the hob spindle can be tilted about its axis and set at any desired angle. This facility is necessary in order to cut helical gears.

The hob and work spindles are connected to each other by gearing, change wheels being fitted into this drive (the wheels not being shown in Figure 7.7). Appropriate change wheels are fitted to give the correct gear ratio between the work and hob. This ratio depends upon the number of teeth to be cut in the gear, and the number of starts in the hob.

7.2 Gear generation principles

In Section 7.1 it was stated that two gears and a rack of involute form having the same module and pressure angle will all mesh with each other. This is the basis of the *generating* processes considered of either an involute gear or involute rack profile, constrained to move in the right relationship to the blank such that gear teeth are generated as the metal is machined. The generating principles of each process referred to in Section 7.1 will be considered here.

Planing process. The planing process makes use of a cutter of true involute rack shape, having rake and clearance, which reciprocates across the face of the blank. Figure 7.8 shows a diagram of a rack

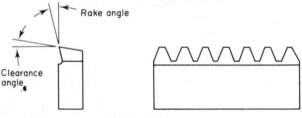

Fig 7.8 Rack cutter

cutter. The cutter and blank are geared together so that the blank rotates in the correct relationship to the longitudinal movement of the cutter, i.e., they roll together as a rack and pinion. This is shown diagrammatically at Figure 7.9.

It will be seen from Figure 7.9 that the tooth height of the rack cutter is twice the tooth dedendum, in order that the cutter touches the root circle and clears the top circle of the gear being cut.

The machining cycle is:

(*a*) Cutter fed into full tooth depth with cutter reciprocating and blank stationary.

(*b*) Blank rotates and cutter feeds longitudinally. An involute shape is generated on the gear teeth flanks by the involute rack cutter.

(*c*) After one or two teeth are completed the blank and cutter stop feeding. The cutter is withdrawn and indexed back to its starting position.

(*d*) Cutter fed to depth and cycle repeated.

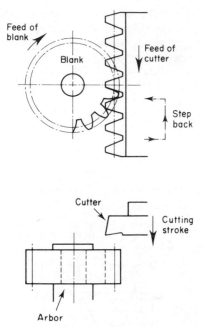

Fig 7.9 Gear planing

The indexing, or 'stepping back' operation described at (*c*) above enables a short rack cutter of a practical length to be used. For faster production, two cutters can be fitted cutting on alternate strokes, thus machining two blanks at one setting. The idle return stroke is eliminated. Gear planing is used for the generation of external spur gears, and where the cutter slide is provided with angular adjustment external helical gears may be cut. If a rack cutter of standard proportions is used for cutting helical gears, then the normal pitch of the helical gear will be slightly incorrect.

The process is ideally suited for cutting large, double helical (herringbone) gears which do not have a clearance recess between the two sets of teeth. A special version of the planing machine is used in this case having two cutter slides inclined at the gear tooth helix angle.

In gear generation by planing, the rack cutter controls the pitch and pressure angle, and the machine gearing controls the number of teeth produced on the gear.

Shaping process. The shaping process makes use of a hardened pinion as a cutter, ground with top rake and clearance. Figure 7.10 shows a diagram of a pinion cutter. The cutter reciprocates across the blank face, and can cut either on the upstroke or downstroke. A relieving

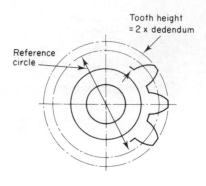

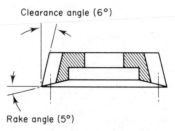

Fig 7.10 Pinion cutter

mechanism enables the cutter to clear the work on the return (non-cutting) stroke. The cutter and blank are geared together so that they rotate in the correct relationship to each other to suit their respective numbers of teeth, i.e., they rotate together as two gears in mesh.

The principle of operation of the process is shown at Figure 7.11. The machining cycle is:

(a) Cutter fed into full depth, with cutter reciprocating and blank stationary. (Blank could be rotating if desired.)

(b) Cutter and blank slowly rotate until all the teeth are generated upon the gear blank.

As with most gear cutting, the blank is usually roughed to say three-quarter depth, followed by a finishing cut. The process can be used for cutting external and internal spur gears, and racks if used with a rack cutting attachment. All these gears can be generated in helical form, but a helical cutter of the correct hand (opposite to the required gear hand) must be used in conjunction with a helical cutter guide of the same angle and hand as the cutter. Usually standard helix angles of 15° and 23° are available, and the process is not recommended for helical (and spiral) gears having a helix angle greater than 35°. Referring to Figure 7.11 when cutting helical gears the cutter is con-

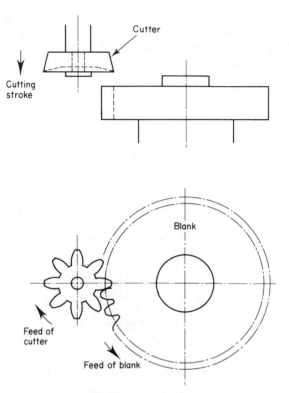

Fig 7.11 Gear shaping

strained to follow a helical path, instead of the vertical, straight line path shown in this diagram. As a special cutter must be used, it is manufactured with the correct normal pitch to suit the helical gear.

This fundamental generating process of work and cutter rotating at constant angular velocities about fixed centres is a form of conjugate machining and gives great versatility. In addition to the range of gears already mentioned, many special shapes such as polygons or cam profiles can be cut. For example a four leaf, clover pattern cutter will generate a square upon the workpiece, if cutter and work rotate at the same velocity.

Hobbing process. This is the fastest of the gear generating processes as it is a continuous cutting process. The reciprocating or indexing movements in the other gear cutting machines are absent from a hobbing machine. This is one of the most fascinating generating processes, and it is difficult to visualise it working until one actually sees the process being carried out.

It may be helpful to first imagine a disc being generated using a slab or roller milling cutter as shown in Figure 7.12.

The milling cutter is rotated at the correct cutting speed to suit the disc material, and is slowly traversed across the face of the disc at the required depth of cut. The disc rotates and will therefore be machined to a reduced diameter. In Figure 7.12 the process is one of up-cut

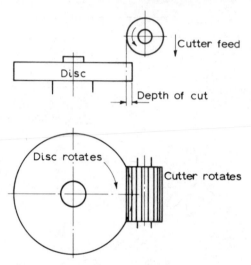

Fig 7.12 Generating a disc by milling

milling, but could equally well be down-cut. Note that there is a distinct relationship between the cutter feed, and speed of disc rotation. If the milling cutter is fed too quickly across the disc face, then a helical groove will be milled on the disc face.

If we wish to generate gear teeth upon the disc, we must replace the milling cutter with a form relieved milling cutter called a hob. This is in effect a worm having a thread of involute tooth rack shape, gashed with form relieved teeth. Figure 7.13 illustrates a gear cutting hob in half-section with the rack profile of the worm thread being shown.

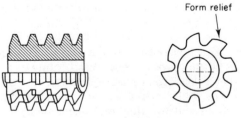

Fig 7.13 Gear hob

The hob is rotated at a suitable cutting speed and fed across the blank face. The hob and blank are geared together so that they rotate in the correct relationship to each other, i.e., they rotate as a worm and worm wheel in mesh. In effect, the worm as it rotates presents an endless set of rack teeth to the blank.

The principle of hobbing is shown at Figure 7.14.

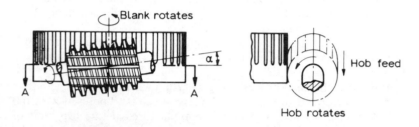

Part section A-A
(True involute rack shape)

Fig 7.14 Gear hobbing

The machining cycle is:

(*a*) Hob and blank rotate.

(*b*) Hob fed into full depth.

(*c*) Hob feeds across blank face and generates involute teeth.

As with all milling-type operations, there must be sufficient clearance on the side of the blank to allow the hob to run out clear of the work. Reference to Figure 7.14 shows one or two important points. The hob must be tilted through its helix angle α when machining straight spur gears, in order to bring the hob teeth into line with the cut gear teeth and avoid interference. Therefore the rack form of the teeth must lie along a line at angle α to the hob axis as shown by section *A.A.* The teeth gashes are usually machined in the hob normal to the tooth helix as shown. Multi-start hobs are used for fast roughing upon production, because a three start hob for example, will cut three gear teeth for one revolution of the hob. The more accurate single start hobs are used for finishing, and one hob revolution generates one gear tooth. Hobbing can give the highest rate of gear production, and the fastest machines are fitted with fully automatic loading and ejecting equipment.

Although we are only concerned with the production of spur gears here, it is interesting to note the versatility of the hobbing process. For example helical and spiral gears can be produced upon a fully universal hobbing machine. A standard hob can be used for cutting helical as well as straight gears, but the hob must be tilted through an additional angle equal to the helical gear helix angle, in order to avoid interference (i.e., if both hob and gear have right-hand helices). In this case, as when using a standard cutter in planing, the helical gear normal pitch will be slightly incorrect. Also, it is necessary to use differential change gears when machining helical gears in order to give an auxiliary rotation to the blank. The hob will still feed down a straight, vertical line as shown in Figure 7.14, but the gear teeth are required to follow a helical line. Hence, the blank must continually accelerate as it were for each increment of hob feed.

Double helical gears can be produced by hobbing if a clearance groove is left between each set of teeth, in order to provide run-out for the hob.

Hobbing is the ideal process for generating worm wheels. The hob is made to correspond to the mating worm which is to be used with the worm wheel in service. The hob is manufactured to the top limit of diameter to allow for the hob teeth being re-ground. (Note that the teeth are reground on their front face, but the hob diameter will decrease after each re-grind due to the form relieved shape of the teeth.) Large worm wheels are usually roughed out using a fly cutter and then finished hobbed. The hobbing process for worm wheels is identical in principle to that used for producing gears, but the hob can be fed to full tooth depth either radially or tangentially.

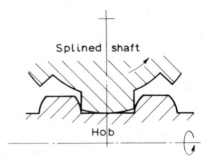

Fig 7.15 Spline hobbing

If the hob tooth profile is made other than involute rack shape, then shapes other than involute can be generated by hobbing. Serrated or splined shafts, for example, can be hobbed and frequently are when required in large quantities. External splines can be more accurately and quickly generated than formed. The required hob tooth profile to produce a standard spline can either be plotted by graphical means on the drawing board, or points on the developed tooth curve can be calculated.

The tooth profile and the resultant spline shape are shown sketched in Figure 7.15.

7.3 Choice of process

In general, the generating processes described in the previous section produce gears faster and more accurately than the forming process. However, each of the generating processes have advantages and limitations when compared to each other, these factors influencing the *choice of the process* for a particular application.

The most important advantages and limitations are given below for each process.

Gear planing.

(*a*) Simple process using a relatively cheap cutter which is easy to regrind.

(*b*) Ideal process for cutting large gears.

(*c*) The only practical method of producing large, double-helical (herringbone) gears, which do not have a clearance groove.

(*d*) Can cut gears up to a shoulder.

(*e*) Discontinuous process due to reciprocating cutting action and stepping-back operation.

(*f*) Cannot produce internal gears.

Gear shaping

(*a*) This is the most versatile of the gear cutting processes being capable of producing internal and external gears, and racks of both straight and helical form.

(*b*) The conjugate generating action allows a great variety of internal and external geometric shapes to be produced.

(*c*) Can cut gears up to a shoulder.

(*d*) A fast process although it is not completely continuous due to the reciprocating action of the cutter.

(*e*) Requires separate cutter and guide for helical gears of a particular hand.

Gear hobbing

(*a*) The fastest of the gear generating processes as it is a completely continuous process.

(*b*) A versatile process as it can produce both straight and helical gears. It is the ideal method for producing worm-wheels in quantity. Can generate shapes other than involute, and is favoured for the production of splined shafts.

(*c*) Produces very accurate gears at high speed.

(*d*) Cannot cut internal gears.

(*e*) Needs clearance to allow for the hob run-out.

(*f*) A hob is a relatively complicated and expensive cutting tool, requiring careful re-grinding.

Learning Objectives 7

The learning objectives of this section of the TEC unit are:

7. Understands the principles and appraises the applications of gear cutting.

7.1 Explains the operating principles of the following methods of spur gear manufacture: shaping, planing, hobbing.

7.2 Contrasts the cutter form and the relative movements of cutter and blank associated with each process in 7.1.

7.3 Identifies the factors which influence the choice of each process.

Exercises 7

1   A gear of 5 mm module has 45 teeth. Calculate: (*a*) reference diameter, (*b*) addendum, (*c*) dedendum, (*d*) tooth working height. (Ans. (*a*) 225 mm, (*b*) 5 mm, (*c*) 6·25 mm, (*d*) 11·25 mm.

2   With the aid of diagrams, describe the operating principles of gear shaping, planing and hobbing machines respectively. The diagrams should clearly show the 'primitive' gearing arrangement between cutter and work.

3   Sketch the cutters used in the following gear cutting processes: (*a*) shaping, (*b*) planing, (*c*) hobbing. Show in diagrammatic form the relative movements of cutter and blank necessary for gear tooth generation.

4   List and describe the factors which influence the choice of a particular process to be used for a gear generation production process.

5   Describe and compare the process of gear tooth forming with that of gear tooth generation. State the advantages and limitations of each.

Chapter 8
Numerically Controlled Machine Tools

Ever since machine tools have been used as a means of quantity production in the engineering industry, the trend has been towards developing semi-automatic or fully automatic machines This is in order to reduce labour costs which have increasingly become the largest single cost factor in the total cost of a product. Techniques have varied in this process of automation, but numerically controlled (*NC*) machines have probably made the greatest impact of all systems in industry.

Development of *NC* systems is still proceeding at a fast pace. The earlier systems (most of which are still in use) consisted of purpose-built control units permanently connected to machine tools. These 'dedicated' or 'hard-wired' systems are termed *NC* machines. Clearly they are relatively inflexible as they are special-purpose machine tools. Recent developments in the electronics industry particularly in the areas of miniaturisation and integration of circuits, has led to the introduction of new, small and powerful computers which can be used to control machine tools in place of the conventional dedicated controller. These newer control systems are known under the general term Computer Numerical Control (*CNC*). To the user the control system 'hardware' (i.e. the component parts which make up the control system) is in principle the same. The advantages of *CNC* are related to the control system 'software' (i.e., the information fed to, and stored in the control system) which allows a great degree of flexibility not obtainable with a *NC* system. It is possible to extend a numerical control system such that using a single computer and data transmission lines, several machine tools can be controlled. These systems are known under the general term Direct Numerical Control (*DNC*).

Technology is now proceeding so quickly that no doubt more significant developments will continue in the field of numerical control in the near future. However, this chapter is essentially an introductory text so no attempt will be made to distinguish between various systems, but rather basic principles only will be examined.

8.1 *NC* operating principles

Any *machine tool control system*, no matter how large or small, is either an *open loop*, or a *closed loop* system.

Open loop control system. If an input signal or command is given to a machine slide to move to a certain position, the slide will move to its ordered position and then stop. If the loop is open, then there is no feed back and the control system has generated no return signal which will indicate that the slide has moved to the correct position. The control loop has not been closed; there is no indication that command and result match, and consequently no means of knowing that there is an error where one exists. See Figure 8.1 which shows an open loop system for the control of a machine slide.

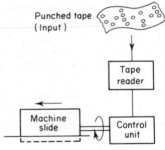

Fig 8.1 Open loop control system

Closed loop control system. In this case there is feed back built into the system. Hence, as the slide moves to its new position as the result of a command signal, its position is monitored automatically. A signal is fed back to the control unit which indicates that the slide has either moved correctly to its new position, or it has not. If not, its position is automatically corrected until it is in the right position. In general, a closed loop system will be more expensive to apply to machine tool control than an open loop system. See Figure 8.2 which shows a closed loop system for the control of a machine slide.

A cam operated automatic lathe is an example of an open loop control system. Here the tools are traversed to their required position under the action of cams. In effect a tool takes up a position as a result of a command signal from a cam, but there is no feed back to indicate that the tool has reached its correct position.

The fully automatic centreless grinding system illustrated at Figure 4.9 is an example of a closed loop control system. Here a wheel head slide positioning unit automatically feeds back impulses so that the position of the wheel slide (which controls the workpiece size) is being continuously monitored.

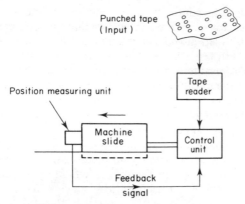

Fig 8.2 Closed loop control system

A closed loop system is said to be 'error-actuated', i.e., the value of the actual slide position is automatically compared with the value of the desired slide position, and an error exists if the two values do not agree; corrective action will automatically follow until the error is eliminated. If the error is zero, no corrective action takes place.

NC **system.** In this system, the programme from which the command signals are derived is stored in numbers, hence the term *numerical control*. This method is ideally suited for the small batch production of parts, particularly very complex parts which require great skill on the part of a machine operator. It is also suited for larger batch production, and the tapes can be stored so that the machine can be quickly set up at any time and the same programme used. Consistency and elimination of error are natural advantages which accrue. In order to help understand a numerically controlled machine tool, we will look first at an operator controlled machine tool (see Figure 8.3) and then see how the human controlled operating functions are replaced by numerically controlled operating functions (see Figure 8.4).

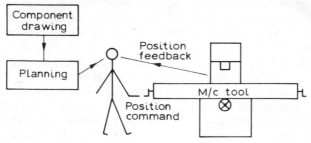

Fig 8.3 Operator controlled machine tool

With an operator controlled machine tool some initial planning takes place before the drawing and material are passed to the operator. The information is fed into the machine via hand-wheels, switches, etc. The operator interprets the component drawing in linear and angular measures which are acceptable to him and the machine. The operator monitors the cutter position during machining and makes any neces-

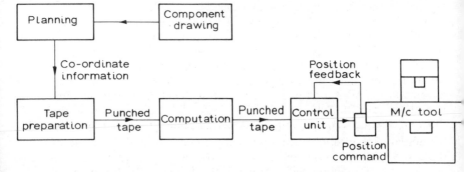

Fig 8.4 Numerically controlled machine tool

sary corrections to ensure suitable output. The system is therefore closed loop with the operator carrying out data processing, position input, position feedback and compensation functions. With a *NC* machine tool various functions carried out by the operator are now undertaken by other means. The machine tool is equipped with a control unit which feeds the position command information to slideway transmission elements and compares this position command with a position feed back signal derived from automatic monitoring of the machine tool slide position. The component drawing must be translated into a form acceptable to the control unit. This is the data processing part of the system where the co-ordinate information is recorded on a tape by means of a teleprinter. Again, the system is closed loop because of the feed back link.

The human operator in addition to controlling the machine has the ability to draw conclusions from the machine behaviour and act accordingly. He can compensate (through experience) for backlash in the lead screw, slide friction, lack of stiffness, etc., and approach the final cut gradually before final commitment to it. The numerical controller on the other hand is intolerant of these characteristics and can only accept an ideal concept. The first effect of fitting numerical control therefore is to decrease accuracy of working (between 0·025 mm and 0·075 mm) compared with that which can be attained by a skilled and careful operator.

Certain aspects of machine design assume greater importance if satisfactory results are to be achieved, and great improvements have been made in overcoming friction problems with re-circulating ball lead screw and nut for example. Note that monitoring devices which feed back information only give positional information about the table (to which workpiece is fastened) relative to the slides, not about the cutting edge relative to the workpiece. The positional measuring (monitoring) devices in general usage are more than accurate enough when the above limitations are considered.

A *NC* system does not only control the machine slide(s) displacement but other functions, such as start, stop or reverse spindle, select desired speeds and feeds, start or stop cutting fluid flow, lock or unlock slides, etc. A fully automatic *NC* machine will have all the machine functions controlled by means of a tape. A less expensive, semi-automatic *NC* machine will have some of the functions under the control of an operator, e.g., on a single-spindle *NC* drilling machine the different cutting tools required for the machining cycle might be changed by the operator.

8.2 The binary numerical system

The *binary numbering system* is used in numerical control systems, but let us first look at the familiar denary numerical system used for our everyday arithmetic. This contains 10 digits which are 0, 1, 2, 3, 4, 5, 6, 7, 8 and 9 respectively, and any number used in the denary system is based upon powers of 10.

e.g., 3406 is made up of:

$$(3 \times 10^3) + (4 \times 10^2) + (0 \times 10^1) + (6 \times 10^0)$$
$$= 3000 + 400 + 0 + 6$$
$$= 3406$$

The binary numerical system, on the other hand, contains two digits which are 0 and 1 respectively, and any number used in the binary system is based upon powers of 2.

e.g., 1011 is made up of:

$$(1 \times 2^3) + (0 \times 2^2) + (1 \times 2^1) + (1 \times 2^0)$$
$$= 8 + 0 + 2 + 1$$
$$= 11$$

Hence, the number 11 in the denary system can be represented by 1011 (one-nought-one-one, not one thousand and eleven) in the binary system.

Some binary numbers and their denary equivalents are given in the table shown below.

Binary number	Denary equivalent	Derivation
0	0	(0×2^0)
1	1	(1×2^0)
10	2	$(1 \times 2^1) + (0 \times 2^0)$
11	3	$(1 \times 2^1) + (1 \times 2^0)$
100	4	$(1 \times 2^2) + (0 \times 2^1) + (0 \times 2^0)$
111	7	$(1 \times 2^2) + (1 \times 2^1) + (1 \times 2^0)$
10000	16	
10101	21	

The last two derivations are left as an exercise.

When comparing the two systems, the binary numbers appear clumsy. For example the denary number 128 (three digits) is represented by 10000000 (eight digits) in binary form. The point is of course, that any binary number can never have more than two types of digits, i.e., 0 and 1. To the electronics engineer this represents the simplest way of representing a number by means of an electrical impulse. Either a hole in a tape allows electrical contact to be made (ON), or the absence of a hole in the tape does not allow electrical contact to be made (OFF). Therefore, a hole can represent 1, and no hole can represent 0, hence a tape can be used to transmit any required set of numbers in binary form, and relatively simple circuitry (highly complicated to most of us) can be used in the control system.

It is appropriate at this stage to consider how this binary numerical system is applied to tape coding.

Tapes. In most numerical control machining systems, some form of tape is used for transmitting information given by the drawing to the computer, and used again to transmit data given by the computer to the machine. Two types of tape are in common use; (*a*) Punched paper tape, and (*b*) Magnetic tape.

(*a*) *Punched paper tape.* This gives a cheap and convenient means of carrying out the first part of the exercise just described, i.e., transmitting drawing information to the computer. A punched hole can represent 1 and the absence of a hole can represent 0. In use the tape is fed over brush contacts which 'sense' the hole pattern and can detect the presence of a hole or not. The brushes close the circuit through a punched hole and trigger off signal currents which cause the appropriate parts of the computer circuit to be actuated. Figure 8.5 illustrates the principle.

In addition to electro-mechanical tape readers working on the principle described, photo-electrical tape readers are used in which the tape is scanned by concentrated light rays falling upon it which detect the presence of a hole or otherwise. This type of tape reader operates at considerably faster speeds than others.

A further alternative type of tape reader is the pneumatic type in which the hole pattern on the tape is sensed by compressed air passing through jets. This is a slower system than others but is adequate for point-to-point *NC* applications (see Section 8.3).

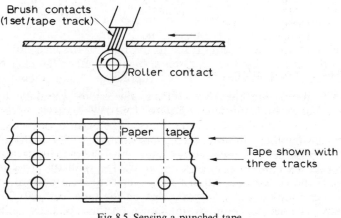

Fig 8.5 Sensing a punched tape

(*b*) *Magnetic tape.* This is usually 6 or 25 mm wide and is exactly like that used in home tape recorders. The principle is similar to that used upon a punched tape, but instead of a punched hole (1) and no punched hole (0) magnetised spots on the tape surface are used. The polarisation of the magnetic spot (either + or −) indicates the binary digit (either 1 or 0).

The feeler sensing pins of the punched tape are replaced by an electric reading head which contains an electro-magnet. See Figure 8.6.

Each magnetised spot on the tape which passes under the reading head will generate an electric impulse in the coil, the polarisation of the spot enabling the system to recognise 1 or 0. The big advantage of

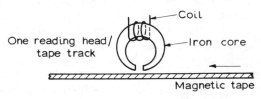

Fig 8.6 Sensing a magnetic tape

magnetic tapes is that far more information can be recorded upon them, and they can be 'read' at a speed faster than is possible with punched tape (or cards), if so desired. It is also more durable but more expensive than paper tape.

It was stated earlier that binary numbers are clumsy compared with denary numbers because of the greater number of digits required to represent a simple number. This is so, and if a true binary numbering system was used to represent one number across a row, then tapes would have to have an unwieldy number of tracks of holes (or spots) making the tape very wide.

In practice however, a modified system of binary numbers is employed, and several rows are used to represent a number. This is shown in Figure 8.7 applied to an 8-track tape.

The first four tracks only (three to the right of the feed holes, and one to the left) are used to represent a binary number. Track 1 represents 2^0, track 2 represents 2^1, track 3 represents 2^2 and track 4 represents 2^3. Hence, across a row of tape, punched holes (or spots) placed in the appropriate position can represent any denary digit up to 9 in value.

The pattern of holes shown in Figure 8.7 represents the value 12·793 6. Any required value can be represented in the system shown by a block of six rows incorporating four tracks. With the direction of tape feed shown the hole pattern should be read from bottom to top. Therefore, numbers can be represented on the tape from 00·000 0 to 99·999 9. The decimal point is not shown on the tape because it usually occurs after two digits when using imperial units, or after four digits

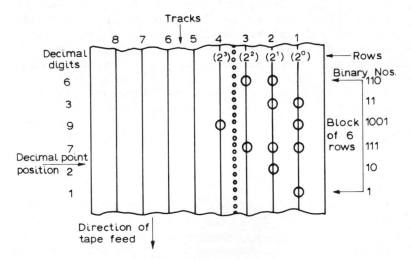

Fig 8.7 Eight-track punched tape

when using metric units. Whether an imperial unit or metric unit machine is being used, the position of the decimal point is fixed and it is therefore unnecessary to use it when writing out programmes (see Section 8.4).

The fifth track (in some systems) is used for odd parity, i.e., a hole is punched or omitted from this track such that every row in the complete tape adds up to an odd number of holes. This is a simple checking device to ensure that the tape punching and tape reading devices are functioning correctly. Should a row appear on the tape with an even number of holes, the machine using the tape will stop. The equipment must then be checked for malfunctioning.

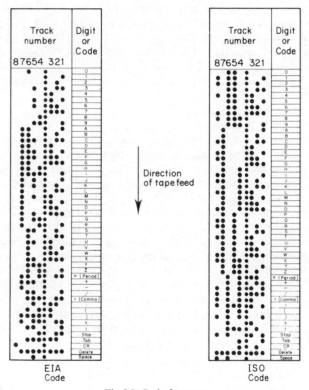

Fig 8.8 Code formats

The sixth and seventh tracks are used for what are known as alpha characters, i.e., alphabetical characters. These are required for coded instructions which will be explained later in the next section.

The eighth track is sometimes used to indicate the end of a block of information, i.e., one hole punched in the appropriate place in the eighth track would indicate the end of a particular sequence of information.

Two systems of codes are used in practice which are based in principle upon the system described above, but vary in matters of detail. The codes are:

(i) BS 3635: Part 1: 1972 (Control input data) code, based upon the ISO format;

(ii) *EIA* (Electrical Industries Association) code, developed in the United States.

The latter code was widely used on the early *NC* machines, but the former is now often used because it is compatible with punched-tape computer systems. The two code formats are shown at Figure 8.8.

Comparison of the two codes at Figure 8.8 shows that the first four tracks used for numeric characters are identical. Track 5 is used in *EIA* code for odd parity, but track 8 is used in the *ISO* code for even parity, i.e., a hole is punched in track 8 when it is required to bring the holes in the row to an even number. The remaining tracks are used differently for the alpha characters.

8.3 Point-to-point positioning

There are a variety of *NC* systems in use, and the great majority are used on metal-cutting machine tools, particularly milling, drilling, boring and turning machines. *NC* grinding machines and machining centres (automatic, multi-purpose machine tools) are also used. A minority area of *NC* application is on flame-cutting machines, turret presses, electron-beam welding machines, etc.

Whatever the type of *NC* machine in use, it will have the capability of carrying out one or more of the following processes: (*a*) Positioning (*P*), (*b*) Line motion (*L*), (*c*) Contouring (*C*), in any one or more of three axes (*X*, *Y* or *Z*).

It has been found convenient to use the abbreviations shown in order to classify *NC* machine tools, as the following examples show: A *NC* vertical drilling machine with positional control to the table axes, and controlled linear motion on the vertical spindle feed would be classed 2*PL*; a *NC* vertical milling machine with contouring facilities in all three axes would be classed 3*C*; and a *NC* lathe with contouring facilities in two axes would be classed 2*C*, for example.

However, there are basically two types of *NC* systems as applied to machine tools:

(i) that in which the finished workpiece or part shape is determined by the controlled relative movements of cutting tool and part during machining. This type of system is referred to as a *continuous path system*;

(ii) that in which the control system determines the relative position of cutting tool and part before the machining operation takes place. This type of system is referred to as a *point-to-point* (or positioning) *system*. This is essentially the simpler system, the basic operating principles of which will be described in this section.

The block diagram shown at Figure 8.9 illustrates a simple point-to-point system, and the steps involved in the process from the com-

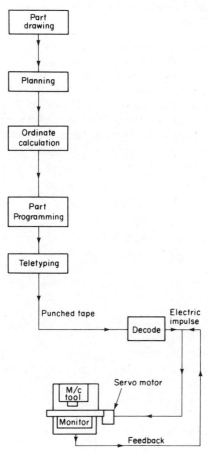

Fig 8.9 Point-to-point *NC* system

ponent (part) drawing to the machining of the part on the machine tool. Each feature will be considered in turn, and it will be assumed that a punched tape is being used.

Part drawing. The object of a point-to-point *NC* process is to successively move the machine tool table (and hence the work) to programmed positions where the table stops and machining takes place. Hence the part drawing ideally should be dimensioned in such a way that all stopping points are dimensioned in rectangular coordinates from a suitable datum point. In the case of a drilling operation the datum point will be the start point of the drill axis, and the stopping points will be the centre parts of the holes to be drilled. Figure 8.10 shows a part drawing of a component to be drilled on a *NC* point-to-point vertical drilling machine. Figure 8.10(a) shows the required hole positions, and Figure 8.10(b) shows the part hole centres

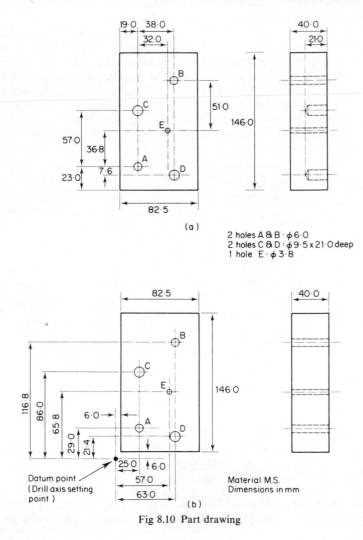

Fig 8.10 Part drawing

as dimensioned for the *NC* process. The component is to be positioned with its edges 6 mm away from the datum point.

The system illustrated at Figure 8.10(*b*) in which all the positional dimensions are stated with reference to a datum point is known as an 'absolute' system, and is normally employed in *NC* point-to-point processing. This is as opposed to an 'incremental' system, in which positional dimensions are taken from previous positions, rather in the manner of Figure 8.10(*a*).

Planning. At this stage the method of manufacture of the part must be planned. The following steps are involved:
  (i) Planning the machine to be used and the method of manufacture. In the case of the part shown at Figure 8.10 the first hole to be drilled is that nearest to the datum point, i.e., hole *A*. The second is the same diameter hole to enable the same drill to be used, i.e., hole *B*. The same sensible pattern of drilling is used to complete holes *C*, *D* and *E*. For the sake of simplicity we will assume that the axis of the machine spindle is positioned on the datum point at the start and completion of the drilling operation, although this is not always so.
 (ii) Planning the cutting tools, drills, jigs or fixtures, and setting-up procedures to be used. A relatively simple fixture is usually required in which the part is located and clamped. See Figure 8.11

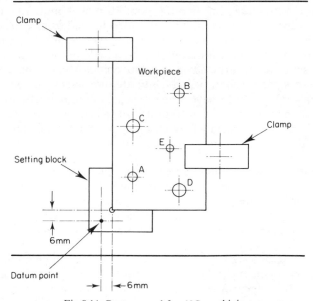

Fig 8.11 Part secured for *NC* machining

which refers to the positioning and clamping of the part shown at Figure 8.10.

(iii) Deciding upon the cutting speeds and feeds.

Ordinate calculation. The rectangular ordinates for the point-to-point movements must be calculated for the appropriate axes of movement, i.e., X, Y or Z. In the case of a simple part, like that at Figure 8.10(b) the part drawing may already contain the desired dimensions. In the case of a more complex part, particularly where holes are irregularly spaced around a pitch circle using polar ordinate dimensions, the use of a computer may be required.

Part programming. All the above planning and ordinate information is written on to a part-programme or 'manuscript', in a manner to suit the system being used. There are different systems of programming in use, and this important stage of the NC preparation process will be considered in more detail at Section 8.4. The part-programme allows the essential planning and ordinate information to be transferred onto a tape, which is the input medium to which the NC machine will respond.

Programming is possible on many NC machines using the 'dial-in' method in which hand dials may be used which are positioned conveniently in the control unit. If settings are required in 0·01 mm increments for point-to-point positioning then five decade (or setting) dials could be used representing say 100 mm, 10 mm, 1 mm, 0·1 mm and 0·01 mm respectively. Each dial is calibrated to read from 0 to 9, hence any settings can be made from 000·00 mm to 999·99 mm. There are one set of dials to control the longitudinal movement of the machine table, and one set to control the crossways movement. Hence, the table can be made to move through two co-ordinate movements for one setting of the dials. They would have to be re-set, to automatically move the table to its next position. Continuous automatic re-setting of the decade dials can be arranged where quantity production of a component is required.

Teletyping. A teletypewriter (teleprinter), operated by a typist to the rules of the code, is used to punch out the pattern of holes on the tape from the information given on the programme. In other words, the tape when produced, will have the identical information punched on it in binary form, to that contained in the programme. A teleprinter is shown at Figure 8.12.

The keyboard has keys for the numerals and alpha characters used in the code. As the typist presses the appropriate key, the correct hole pattern will automatically be punched upon the paper tape issuing from the teleprinter. Also, a check-out sheet will be fed from the front

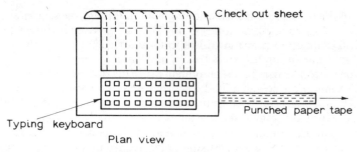

Fig. 8.12 Teletypewriter

of the machine which the typist can read, and check, that she is copying the programme correctly.

Decode. The punched tape is mounted on a play-back tape deck built into a control unit which is adjacent to the machine tool. This tape deck has a reading head which decodes the binary information on the tape as it feeds past it. This information is passed from the reading head to the control unit in digital form as a series of electrical pulses. The tape deck speed has a controlling influence upon the ultimate machine slide velocity along its slideway. Standard tape speeds of 381 mm/s, 190·5 mm/s and 95·25 mm/s are used.

Finally, electric impulses or signals are fed by the control unit to the servo motors which control the machine table (slide) movement and the cutter head. These command signals generated by the control unit are a function of the pulses received from the tape, but magnified in order to actuate the machine control units. All the operator has to do initially is to position the workpiece in the fixture, set the cutter to the datum point, then press the start button. Machining will then commence, and all the operations specified on the original programme will be carried out automatically until the cycle is complete.

Servo motor. The servo motor receives and acts upon the series of electric command signals from the control unit. The motor response must be accurate to drive the work against the cutting forces.

Servo motors are usually of two types: (*a*) high frequency electric, or (*b*) hydraulic. Let us briefly examine each.

(*a*) *High frequency electric servo motors.* These are similar in operation to the normal type of induction motors used upon machine tools at a frequency of 50 Hz. However, the servo motor is operated on a three-phase supply at 400 Hz giving full power at 167rev/s. Consequently, the maximum power available will be limited and is usually only to the order of 0·55 kW. This is a disadvantage where reasonably heavy cutting is required.

(b) *Hydraulic servo motors.* A rotary hydraulic motor is in effect the opposite of a rotary hydraulic pump. The pump rotor is driven by an electric motor, the rotation of the rotor drawing fluid into the pump and expelling it under pressure. In the case of the motor, fluid is pumped into the motor under pressure causing the rotor to turn. Hydraulic motors are used of 2·25 k W with an accuracy of response of 0·005 mm, up to 11·25 k W giving an accuracy of response of 0·025 mm.

The hydraulic fluid is metered to the motor rotor by means of an electro-hydraulic transducer which receives its command signal from the control unit. (Note. A transducer is a device used for converting a signal of one kind into a corresponding physical quantity of another kind, such as a force or displacement. A resistance strain gauge is a transducer, because a change in the length of the gauge can be related to a change in the electrical resistance of the gauge.)

A few modern systems make use of a hydraulic ram displacement, rather than a servo motor.

Monitor. A feed-back system is used on *NC* machines to monitor the slide position when a closed-loop system is employed.

The monitoring systems used for numerical control machine tools are, in effect, position transducers, which enable feed back signals relating to the table (and hence work) position to be transmitted to the control unit. All systems used are either (a) analogue of (b) digital in principle.

(a) *Analogue.* This is the term used to refer to a quantity which resembles in certain respects, another quantity. A slide rule is an analogue device, where a length on a scale is analogous (not equal to) to a number, i.e., the value 4 on the scale might be 150 mm away from the zero mark on the scale, therefore 4 can be said to be analogous to 150 mm. In analogue monitoring devices, an electrical unit such as voltage might be related to the position of work relative to the cutter axis, in millimetres, i.e., 10 volts, for example, could indicate 250 mm of movement of the work. All analogue feedback systems allow rapid indexing from one position to another and are ideal for point to point positioning systems.

(b) *Digital.* The amount of movement of the worktable is measured in discrete (separate) quantities. The number of discrete quantities, say of 0·025-mm magnitude are counted by some device included in the monitoring system. These quantities, or digits, are counted at very high speeds.

Three factors are important in any monitoring, or measuring system, these being:

Accuracy, which is the accordance of the measurement with the measured quantity.

Resolution, which is the smallest value to which the measurement will respond.

Repeatability, which is the ability of the system to repeat a given input value.

There are many monitoring devices in use, each of which possess these three factors to a greater or lesser degree. However, there will only be space to consider two types of monitoring systems, these being suitable for point-to-point systems.

They are:

(i) Rotary type Synchro, and

(ii) Optical gratings.

(i) *Rotary Type Synchro.* This is an analogue measuring system, which is relatively simple and cheap. It measures rotation of the lead screw, which can be related to table movement. See Figure 8.13.

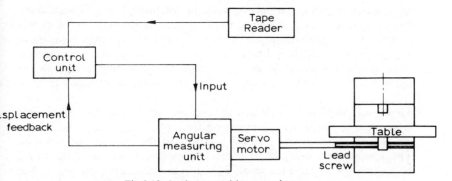

Fig 8.13 Analogue position transducer

Lead screw rotation is measured by means of scanning a potential transducer, the voltage picked off being proportional to the rotation from a machine datum. Disadvantage is that the lead screw rotation is being measured and not the work (table) displacement, therefore variations due to mechanical linkage, and in particular backlash, are not accounted for.

This system has high accuracy if steps are taken to overcome the problems of backlash, etc., and is ideal for point-to-point machining.

(ii) *Optical Gratings.* This system may be analogue, but if combined with a pulse counting device, can be used as a digital measuring system. Although relatively expensive, it is a high-class monitoring system and is worth examining in some detail. Figure 8.14 is a diagram of the main features.

An optical scale or grating is fixed to the machine bed, and super-

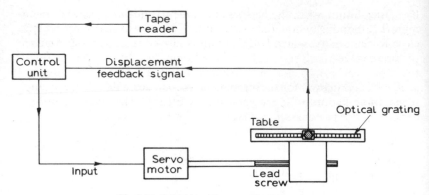

Fig 8.14 Optical grating monitoring system

imposed above it is a short length of scale (about 19 mm diameter) fixed to the table. This small index scale is rotated at an angle to the long scale and scans the bottom scale as the table moves. These transparent gratings have surfaces which bear a large number of evenly spaced lines as grooves (say 100/mm). The grooves are impressed on to a gelatine coated glass blank, the glass acting as a backing. The grooves are too close to be numbered, but can be counted as the table moves by using two scales as described to produce moiré fringes.

Theory of moiré fringes. When two scales are superimposed and one is inclined to the other, illumination from the back will reveal fringes which run approximately at right angles to the lines of the gratings. The spacing (wavelength) of these fringes is a function of the angle between the lines, and the smaller the angle, the larger the spacing. See Figure 8.15.

If the top grating is moved along as shown at right angles to the lines of the fixed grating, then a fringe will move to a position occupied

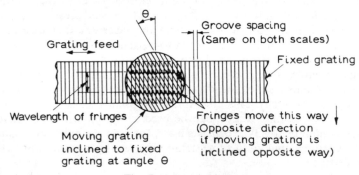

Fig 8.15 Moiré fringes

by its neighbour when the top grating has traversed one groove spacing. If a limiting slot equal in width to half fringe wavelength is placed close to the top grating, as one grating is moved relative to the other, the passage of the fringes causes the illumination at the slot to fluctuate between brightness and darkness. Using a light sensitive device such as a photo-cell the movement of the fringes can be electrically recorded and counted as shown in Figure 8.16. The gratings are not in contact hence no wear takes place.

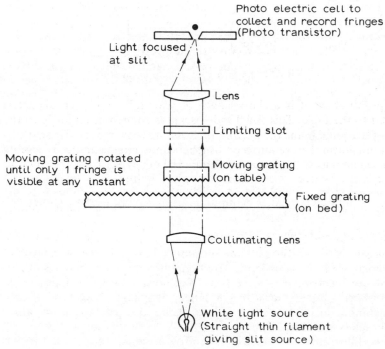

Fig 8.16 Optical system

Referring to Figure 8.16, the lens which is above the limiting slot collects the diffracted light (which need not be monochromatic) and focuses it at the slit. This slit is arranged to select one light of a particular wavelength from the spectra of white light, to suit the photo-cell. These types of optical gratings are sometimes called diffraction gratings because they make use of diffracted light. The latest technique is to use etched stainless steel gratings which reflect light instead of transmitting it. (Note: Monochromatic light is light of a single wavelength which requires a special light source. Diffracted light is light which is broken up into its constituent parts and is a phenomenon which can be caused by passing light through a narrow slit.)

A pair of coarse gratings (4 lines/mm) readily demonstrate moiré fringes, but fine gratings used in practice require an optical system described above. The intensity of the light varies sinusoidally as one grating moves past the other, and a photo-cell placed behind the exit slit will produce an *AC* current, each cycle of which indicates a movement of one grating division. The current is used to actuate an electronic pulse counter which simply registers the cycles of the current and thus counts the number of complete spaces moved by the grating. Therefore 100 lines/mm on the grating gives a resolution of 0·01 mm. An individual counter is used for the control of each machine axis, and these may be zeroed at any convenient slide position. Hence, each slide displacement is measured relative to the selected zero position. Finally, a comparator unit is used to compare the actual slide position with the desired slide position, where necessary automatic corrective action taking place. This is an 'error-actuated' system as described in Section 8.1.

8.4 Part programme

Programming for a point-to-point type of *NC* operation can be carried out using one of the following three methods:

(i) 'Dial-in' method (briefly considered in Section 8.3);
(ii) Computer-aided programming (time saving where complex calculations are required);
(iii) Manual programming.

Again, a choice of part-programming languages are available. A language is a set of coded instructions in a form which the *NC* control unit can accept. As we have seen, ordinate dimensions can be translated onto a tape using a binary language or code. Similarly, machining instructions can be translated onto a tape using alpha characters in coded form.

To illustrate the process of part programming we will use the part shown at Figure 8.17 as an example. It will be assumed that a manual programming system is being used to the *EIA* code, and a punched paper tape is the controlling medium. The machine to be used is a 2*PL* designated vertical drilling machine with an automatically indexing turret as shown at Figure 8.18.

In the interests of simplicity only the main programming features will be referred to, and matters of detail (although important in practice) will be ommitted. The four holes in the part shown at Figure 8.17 are to be completed by a *NC* machine, and the relevant ordinate and planning information is given at Figure 8.19. The datum is positioned 150 mm from the component centre measured on the *X* and *Y* ordinates. Assume the spindle speed to be constant at 650 rev/min., and the feed rate to be constant at 0·15 mm/rev. (97·5 mm/min.).

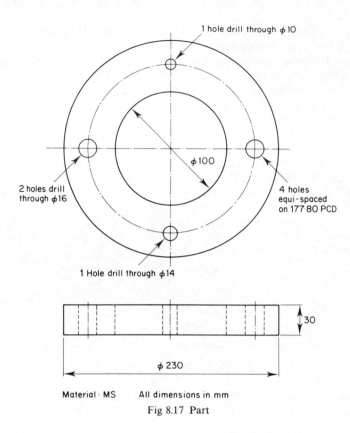

1 hole drill through φ 10

φ 100

2 holes drill through φ16

4 holes equi-spaced on 177·80 PCD

1 Hole drill through φ14

30

φ 230

Material : MS   All dimensions in mm

Fig 8.17 Part

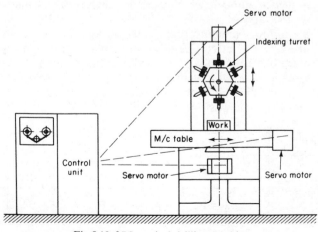

Servo motor

Indexing turret

Work

M/c table

Control unit

Servo motor

Servo motor

Fig 8.18 2*PL* vertical drilling machine

A part-programme is set out in tabular form, as shown by the column headings at Figure 8.20.

The machining instructions indicated on the part-programme headings at Figure 8.20 are represented on the tape in coded form using the

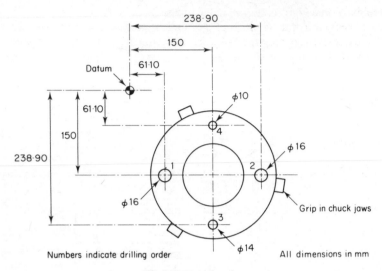

Fig 8.19 Part drawing

alpha characters shown, e.g., F = feed rate. These are briefly explained below:

N = sequence number, which precedes each block of information on the tape;

G = preparatory function represented by a two-digit number prefixed by the letter G and is a command to the machine to carry out a particular standard machining function, e.g., $G79$ = mill cycle, $G81$ = drill cycle, $G84$ = tap cycle, etc.;

X = required slide position in X axis $\longleftrightarrow$;

Y = required slide position in Y axis

Sequence number	Prep. function	Co-ordinates		Feed rate	Spindle speed	Tool number	Miscell. function	End of block
N	G	X	Y	F	S	T	M	(CR)

Fig 8.20 Headings for part-programme

F = feed rate (mm/min.). This is written as a three digit number prefixed by the letter F. The first digit is a decimal multiplier equal to 3 + number of feed rate digits to left of decimal point, and the second and third digits represent the feed rate to two-digit accuracy. Hence, 97·5 mm/min. feed rate has the code $F597$;

S = spindle speed (rev./min.). This is written in coded form in similar fashion to the feed rate, but the three digit number is prefixed by the letter S. Hence, 650 rev./min. spindle speed has the code $S665$;

T = tool number;

M = miscellaneous function number, represented by a two-digit number prefixed by the letter M, and is a command to the machine to carry any necessary function not yet accounted for, e.g., $M03$ = spindle start (CW), $M06$ = tool change, $M10$ = clamp, etc.;

CR = end-of-block character, which indicates the end of a *block* of information. This command stops the tape reader after one block of information has been read, and before the next block is presented. Instructions on a tape are fed in block by block, each block being sub-divided into *words* (each word representing one complete element of information, e.g. $Y015000$), and *characters* (each character being a number or letter represented by a single row of holes across the tape, e.g., M). See Figure 8.22.

A completed part-programme for the hole drilling sequence on the part shown at Figure 8.19 is set out at Figure 8.21.

| N | G | Co-ordinates | | F | S | T | M | (CR) |
		X	Y					
N 001	G 81	X006110	Y015000	F 597	S665	T 01	M 03	*
N 002	G 81	X023890					M 03	*
N 003	G 80						M 06	*
N 004	G 81	X015000	Y023890			T 02	M 03	*
N 005	G 80						M 06	*
N 006	G 81		Y006110			T 03	M 03	*
N 007	G 80						M 30	*

Fig 8.21 Part-programme

The simplified part-programme shown at Figure 8.21 contains the following *EIA* coded instructions:

$G80$ = cancel fixed cycle;
$G81$ = drill cycle;
$T01$ = tool number 1 (φ 16 drill);
$T02$ = tool number 2 (φ 14 drill);
$T03$ = tool number 3 (φ 10 drill);
$M03$ = spindle start (*CW*);
$M06$ = automatic tool change;
$M30$ = end of tape (stops spindle, cutting fluid and feed; resets machine; rewinds tape).

Figure 8.22 illustrates the first block of information on the tape corresponding to sequence number *N001*.

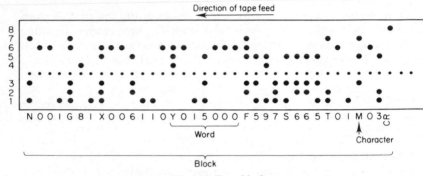

Fig 8.22 Tape block

8.5 Justification of process

The NC process is ideally suited for the batch production of complex parts, which would require great skill on the part of a conventional machine operator. Tapes can be stored so that the *NC* machine can be quickly set up at any time, the same programme easily being repeated. Consistency and elimination of error are *advantages* which naturally accrue.

It should be realised that the capital cost of a *NC* machine can be up to 30–200% higher than that of a conventional machine tool. Further, expensive ancillary equipment is necessary (tape punch for example), and trained personnel will be required. It follows that in order to quickly gain a satisfactory return on the high capital investment, a *NC* machine must be fully utilised. It is usually necessary to introduce shift working in order to gain the maximum exploitation of the equipment.

The main advantages of *NC* machine tools compared to conventional machine tools are as follows:

(i) *Less operator skill required.* Hence, less error and scrap results, and more consistent work is produced.

(ii) *Machine utilisation can be increased.* This is possible because the setting-up time is less, and so the machine is used for metal cutting for a greater proportion of the time available.

(iii) *Less jigs and fixtures needed.* Also the fixtures are usually simpler in design. Therefore, fixture manufacture and storage costs are less.

(iv) *Management control is more accurate.* The *NC* production time/piece is usually accurately known well in advance of the actual production process, i.e., at the programming stage. This is of great value for production and cost control purposes.

As the *NC* process is of an automatic nature, it might be thought that it is most suitable applied to mass production. However, this is not so and it has been found that it is economically viable in the small to medium batch production range, and this is the main field of application. In any event a cost comparison should always be made between *NC* machining and conventional machining when each is a viable alternative, and a break-even analysis is a simple, but effective way of making a cost comparison. (See Section 6.3 for explanation of break-even charts.) The following example will serve to illustrate the principle.

Example 8.1

A component can be produced with equal facility upon either a numerically controlled milling machine, or an operator controlled milling machine by more conventional means.

(a) Which process should be chosen for minimum costs if two components only are required? (Assume no special tooling for the conventional machine and hence no fixed costs, but that it is a toolroom universal milling machine with high overheads.)

(b) Which process should be chosen for minimum costs if a batch of 100 components is required? (Assume special tooling such as a fixture and gauges are required for the conventional machine, which is a plain miller on a production line.)

The following cost information is known.

	(a)		(b)	
	NC machine	Conventional machine	NC machine	Conventional machine
Fixed cost	£80	—	£80	£300
Labour/part	£1·50	£12·50	£1·50	£0·35
Material/part	£1·00	£1·00	£1·00	£1·00
Overheads/part	£3·00	£30·00	£3·00	£2·25

Solution

(a) Variable costs/part for *NC* machine = £1·50 + £1·00 + £3·00 = £5·50.

Variable costs/part for conventional machine = £12·50 + £1·00 + £30·00 = £43·50.

The costs can now be plotted on a break-even chart. (See Figure 8.23)

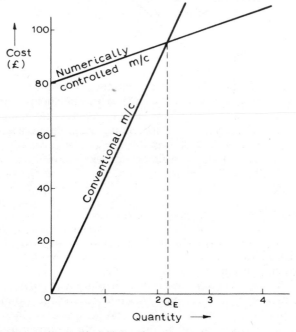

Fig 8.23 Break-even cost chart

Q_E is equal to approximately 2.2 components.

The conventional machine should be chosen if two components only are required.

(b) Variable costs/part for *NC* machine = £5·50, as before.

Variable costs/100 parts for *NC* machine = £550·00

Variable costs/part for conventional machine = £0·35 + £1·00 + £2·25 = £3·60.

Variable costs/100 parts for conventional machine = £360·00.

The costs can now be plotted on a break-even chart (Figure 8.24) to find the value of Q_E.

Q_E is equal to approximately 116 components.

The numerically controlled milling machine should be chosen if 100 components only are required.

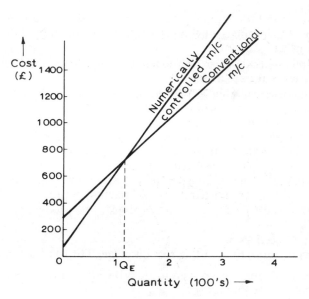

Fig 8.24 Break-even cost chart

8.6 Drawing office practice for *NC* processes

The use of *NC* machine tools in a manufacturing factory will inevitably bring about a change in the conventional *methods* practised in the *drawing office*.

Two obvious changes are:

(i) Components should be designed to take full advantage of *NC* processes. This is particularly true of continuous path systems;

(ii) detail drawings of parts should be dimensioned to suit *NC* processes. With the point-to-point system, dimensioning in co-ordinates as shown at Figure 8.19 is preferable to the conventional dimensioning shown at Figure 8.17.

However, more fundamental changes of organisation may be called for if the most efficient system is to evolve. For example, in a conventional engineering works the functions of design, part detail-draughting and planning are usually separate in principle (even if they are being carried out in the same office). Ideally, for *NC* production, the functions of design, detailing, planning and part-programming should be brought together as an integral process. Drawing/measuring machines are available which enable dimensional information to be automatically transferred from the drawing to the part-programme. Further, such machines are available with a computer facility which enables machining data to be stored; such information in turn can be automatically transferred to the part-programme.

The trend in future, particularly where complex work is being produced on *NC* machines, will be to the use of computerised draughting systems which consist of:

    (*a*) Mini-computer (in which standard data will be stored ready for retrieval at any time);

    (*b*) graphical visual display unit;

    (*c*) digitiser table (on which the design is developed, the result being shown on the visual display unit).

Such a system enables a designer to produce a complete, detailed design scheme with all the information being stored in the computer. The design can then be checked and verified before drawings and part-programmes are finalised. Again, it is clearly only a step away from a control tape being produced as part of the design process hence eliminating the need for conventional drawings to be produced.

Learning Objectives 8

The learning objectives of this section of the TEC unit are:

8. Understands the concept and justifies the application of numerically controlled machine tools.

8.1 Outlines the operating principles of numerically controlled machine tools.

8.2 Explains the binary numerical system and its application to tape coding.

8.3 Explains using a block diagram, a simple point to point system.

8.4 Prepares a programme from a simple component drawing.

8.5 Justifies the use of numerically controlled machines in preference to conventional machines.

8.6 Explains the influence of the use of numerically controlled machines on drawing office practice.

Exercises 8

1   Explain what is meant by the term 'numerically controlled machine tools'.

2   Describe a closed loop and an open loop control system (for machine tools) respectively. Give an example of each.

3   Differentiate between a decimal numerical system and a binary numerical system. Which forms the basis of a numerical control system for macine tools and why?

4   Explain how numerical information can be stored upon (*a*) a punched tape, (*b*) a magnetic tape. What is the advantage of each method?

5   (*a*) Convert the following decimal numbers into equivalent binary numbers: (i) 13, (ii) 28, (iii) 123, (iv) 572.

(*b*) Show the pattern of punched holes on a 4-track tape to represent the value 153·96.

6 With the aid of diagrams, describe a fully automatic point-to-point machining system for drilling components using (*a*) tape control, or (*b*) decade dial control.

7 Show, with the aid of sketches, how the position of the cutter relative to the workpiece can be accurately monitored on an automatically controlled machine tool. Describe two different types of monitoring systems.

8 Draw up a simple part-programme for the operation of drilling the three holes in the part shown at Figure 8.25 to be produced in batches on a point-to-point *NC* machine. Use the *EIA* code.

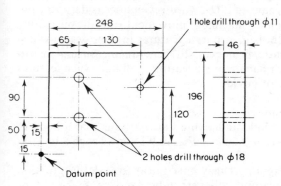

Spindle speed 580 rev/min
Feed rate 0·12 mm/rev

All dimensions in mm

Fig 8.25 Part

9 Compare the use of a *NC* vertical drilling machine to that of a production type multi-spindle drilling machine, for the batch production of drilled components.

10 What do you think is the main influence of the use of *NC* machines on drawing office practice?

Chapter 9
Kinematics applied to Instrument Design

Kinematics is the branch of mechanics relating to problems of motion and position, without regard to the factors of mass and force which come under the term 'dynamics'. (The word kinematics is derived from the Greek word 'Kinema', meaning movement.) Both measuring machines and machine tools incorporate features in their design which are based upon kinematic principles. In general, measuring instruments or machines which have moving features will have such features moving in straight lines or circles.

9.1 Six degrees of freedom
The most important principle underlying kinematics is that a body in space possesses *six degrees of freedom*. Figure 9.1 shows such a body in space.

The body shown at Figure 9.1 may have linear (translatory) motion along any one of the three mutually perpendicular axes X, Y or Z, or may have rotational movement about any one of the three axes respectively. Hence the body has three translational degrees of freedom and three rotational degrees of freedom making a total of six.

The aim of kinematic design is to prevent movement in certain of the degrees of freedom by applying constraints. If six constraints are applied (one for each degree of freedom) then the body will be fixed in space. If five are applied then the body may move backwards or forwards along one axis, or alternatively may rotate about one axis. In instrument design it is nearly always necessary to allow one degree of freedom, i.e., by applying five constraints, or sometimes to completely constrain a body. In any event, in kinematic design it is necessary to identify the degrees of freedom required and to apply constraints to the remainder. Further, only the minimum number of constraints essential for this purpose should be applied, otherwise redundant constraints will be present.

9.2 Kinematic instrument design
The two most common *kinematic design systems* will be considered here, viz., those having:

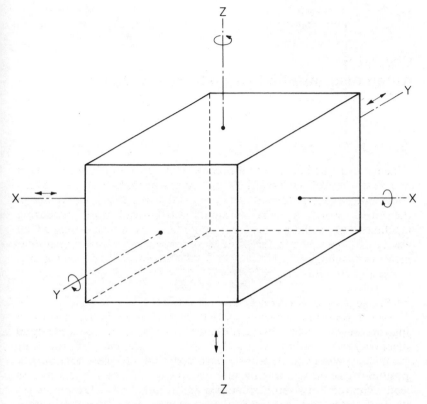

Fig 9.1 Body in space

(a) no degrees of freedom (complete constraint),
(b) one degree of freedom (five constraints).

Other systems having more degrees of freedom are rarely met in measuring instrument design.

(a) *No degrees of freedom*. Consider the case of a simple instrument platform mounted upon a stand such that six constraints are applied. Such a system, sometimes known as a Kelvins Coupling, is illustrated at Figure 9.2.

The platform is supported upon a tripod of three balls which rest in a pyramidal (or trihedral) hole, a vee-groove and plane slot respectively, cut in the stand. One ball foot resting in the three-cornered hole makes contact at three points (the hole preventing all translatory movements). The second ball foot resting in the vee-groove makes contact at two points (rotation about two axes being prevented). The third foot resting on the plane makes contact at one point, the remain-

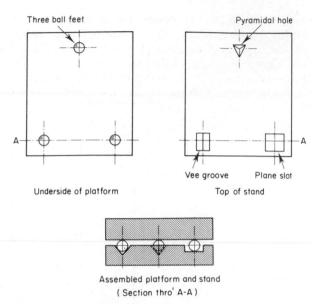

Fig 9.2 Platform and stand

ing rotatory movement being prevented. The only force present is that of gravity, six points of contact being made and the platform being completely constrained. It will be noted that no great accuracy of positioning of the locating parts is necessary, that any thermal dimensional changes between the platform and stand can be accommodated without stress, and that the two parts can be precisely assembled together without difficulty should the platform be removed from the stand. These are important conditions which must be met if the instrument is to function satisfactorily.

(*b*) *One degree of freedom.* This is the most common kinematic principle to be found in measuring instrument design. Consider first a system having only one translatory movement. If a second vee-groove is substituted for the pyramidal hole in the stand shown at Figure 9.2, then the top of the stand will be as shown at Figure 9.3.

When a platform having three ball feet (Figure 9.2) is placed upon this stand, it will be free to move along one translatory axis, i.e., it will have one degree of freedom.

In practice the vee-grooves are made continuous, and rolling balls are substituted for the ball feet in order to reduce friction to a minimum. The arrangement is shown at Figure 9.4.

It will be seen from Figure 9.4 that the stand has two vee-grooves, and the platform has one vee-groove and a flat. The three ball bearings are kept in place by a retaining plate or pegs. The platform makes contact with the balls at five points, hence leaving one translatory degree of freedom along the axis shown. No redundant constraints are present as the minimum number of constraints has been applied, namely five. This type of slideway arrangement is commonly found in measuring instruments, such as a travelling microscope, a thread pitch measuring machine or a floating carriage micrometer as illustrated at Figure 9.5.

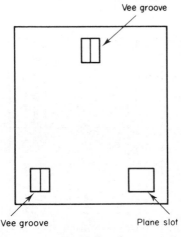

Fig 9.3 Stand

The measuring machine shown at Figure 9.5 consists of three main units. These are a base casting, and a lower and upper carriage respectively. The base has housings to support the work centres. Two vee-grooves in the base lie parallel to the centres and support the lower carriage. This carriage has two conical pegs resting in one vee-groove, the opposite side of the carriage having a flat resting upon a ball which lies in the second vee-groove of the base. The upper surface of the lower carriage also has two vee-grooves which are at right angles to the base vees, the upper carriage being supported in the vee-grooves on ball bearings. This arrangement is as shown at Figure 9.4. Hence, the upper carriage floats freely on the balls, this being the carriage from which the machine derives its name. The system described is truly Kinematic in that freedom of translatory movement is allowed in two directions as shown, which are at right angles to each other.

Next, consider a system having only one rotational movement. This is difficult to achieve with precision on measuring instruments which require rotation only about a single axis, such as optical dividing heads, spectrometer tables or theodolites for example. Taper or conical bearings are usually used, and Figure 9.6 shows a typical design.

The spindle at Figure 9.6 has a conical bearing surface at its upper

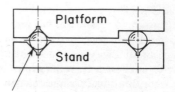

Three ball bearings

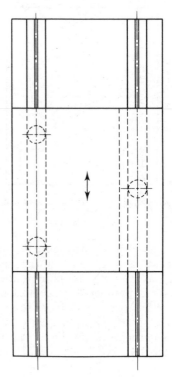

Fig 9.4 Platform and stand

end which rests upon ball bearings. The long stem of the spindle is contained in a plain cylindrical bearing. Theoretically, only one degree of rotational movement is present, but the clearance in the plain bearing allows 'play' in the form of slight translatory movement. A wire clip pressing at three points removes this movement hence leaving one degree of rotational movement. The Kinematic bearing shown has little friction, and is therefore very free running.

To complete this section other examples of good kinematic design applied to instruments are given below with a brief description. They

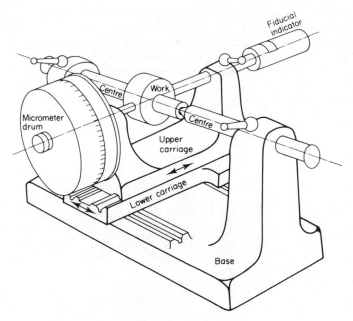

Fig 9.5 Floating carriage micrometer

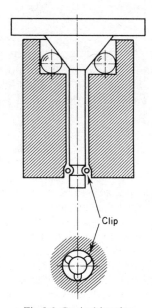

Fig 9.6 Conical bearing

are all based upon the use of metal strips, these normally being made of tempered steel. Strips are used extensively in measuring instruments as couplings, pivots, magnifiers of small movements or parallel motion devices. They do not have the problems of wear, friction and lubrication associated with other types of movable parts.

(i) Crossed strip hinge. A simple hinge of this type is shown at Figure 9.7.

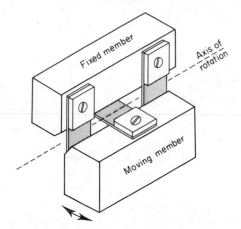

Fig 9.7 Crossed strip hinge

The hinge shown is used as pivot on part of the mechanism of a Sigma Mechanical Comparator, as shown at Figure 9.8.

Movement of the comparator plunger causes a vertical movement of the knife edge shown at Figure 9.8. This in turn causes the arms attached to the moving member of the crossed strip hinge to make an

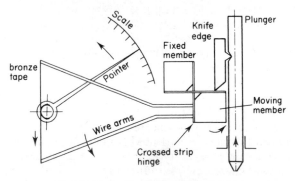

Fig 9.8 Application of crossed strip hinge

angular movement around the axis, leading to the movement of the
pointer against the fixed scale. The radius at which the knife edge acts
from the axis determines the magnification of the comparator.

(ii) Parallel strips. A parallel strip type of support is shown at Figure
9.9.

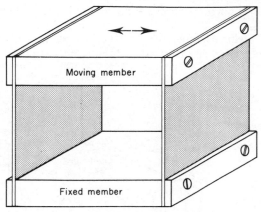

Fig 9.9 Parallel strip support

The type of parallel movement device shown at Figure 9.9 is used as
part of the mechanism of the Eden–Rolt Comparator, which was designed for the calibration of slip gauges. The use of the parallel strip
support is illustrated at Figure 9.10.

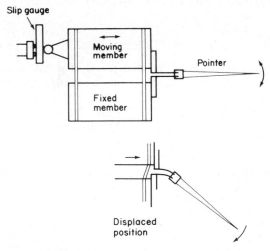

Fig 9.10 Application of parallel strip support

Figure 9.10 shows a slip gauge inserted in the comparator. As the moving member is displaced to the right, it causes the pointer to swing in an arc, as the pointer is attached to both members by strips. A very slight linear displacement of the moving member (caused by the difference in size between the standard gauge and the slip gauge being calibrated), is magnified many times by the rotary movement of the pointer tip.

(iii) Twisted strip magnifier. Figure 9.11 shows a twisted strip magnifier of the type used in the Johanssen Mikrokator type of comparator.

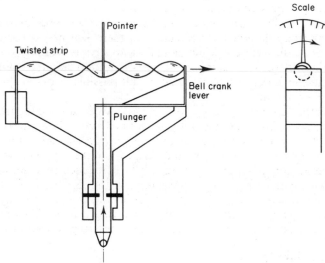

Fig 9.11 Twisted strip magnifier

The twisted strip made from thin metal is permanently deformed by twisting to form two opposite hand helices from the centre to the ends, i.e., the two ends have opposite twists. When the ends are pulled the strip will untwist. The pointer fixed to the centre of the twisted strip will rotate a considerable amount for a small tension applied to its end. This tension is applied by the comparator plunger as shown, when the instrument is used for measurement, and the magnified movement of the pointer takes place over a circular scale.

9.3 Minimisation of errors

In general measuring instruments designed to kinematic principles will have any inherent *errors* reduced to a *minimum*. The following points are worth consideration.

(a) Kinematically designed instruments having a certain degree of accuracy can be manufactured to a lower order of accuracy of construction and assembly. The very highest skills of workmanship are not always necessary in order to produce a high-class measuring instrument to a kinematic design. This is an obvious economic advantage, and is well illustrated by the classic example of the Kelvin coupling shown at Figure 9.2. It will be apparent that a high degree of accuracy of positioning of the holes, slots and balls is not necessary in order to produce an accurate location of the platform on the stand. The platform can be precisely assembled again and again with the minimum of error in the positioning of the two parts.

(b) There will be the minimum number of constraints present in a kinematically designed instrument, i.e., there will be no redundant constraints present which could set up stresses leading to errors.

(c) Contact between component parts is usually at single points in kinematic designs, hence friction and possible errors are reduced to a minimum.

(d) Errors are also reduced to a minimum because a kinematically designed instrument is unlikely to have heavy clamping of parts, will be minimally affected by temperature changes, and inertia will be reduced to a minimum. Because of these particular effects being minimised, errors in turn will be reduced to a minimum.

(e) Kinematic instrument designs, because they are basically simplistic in essence, lend themselves to adjustment during the assembly and testing periods of the manufacturing process. This leads to the most satisfactory and error free operating conditions.

Learning Objectives 9
The learning objectives of this section of the TEC unit are:

9. Understands the principles and appraises the applications of kinematics to instrument design.

9.1 Explains the concept of the six degrees of freedom and minimum constraint.

9.2 Identifies 9.1 in measuring instrument design, e.g., floating carriage micrometer, pitch measuring machine.

9.3 Relates the kinematic design of measuring instruments to the minimisation of errors.

Exercises 9

1 Explain the concept of 'six degrees of freedom'.
2 What is meant by (a) kinematic design, (b) degrees of freedom, (c) minimum and redundant constraints?

3   Sketch and describe a device having no degrees of freedom with minimum constraints.

4   Draw a diagram to illustrate a kinematic instrument design having (*a*) one translational degree of freedom, or (*b*) one rotational degree of freedom. In each case name a measuring instrument which has such a feature.

5   With the aid of diagrams describe (*a*) a crossed strip hinge, (*b*) parallel strips, (*c*) a twisted strip magnifier. In each case name a measuring instrument which has such a feature, and sketch the application.

6   Explain how kinematic design is likely to reduce operating errors in an instrument to a minimum.

7   Examine a floating carriage micrometer of the type shown at Figure 9.5, and comment upon the various aspects of its kinematic design, especially with respect to the arrangement of the slides.

8   Examine a pitch measuring machine of the type shown at Figure 10.7, and comment upon the various aspects of its kinematic design, especially with respect to the slide arrangement, and the stylus head and fiducial indicator.

Chapter 10
Screw Thread Measurement and Gauging

Although Whitworth, BA and Unified thread forms are still in general use, it is likely that they will gradually be replaced by the *ISO* metric thread form. This is shown at Figure 10.1, and this chapter will be confined to consideration of threads of *ISO* form only.

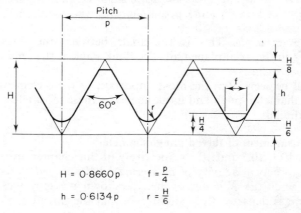

$$H = 0.8660p \qquad f = \frac{p}{4}$$
$$h = 0.6134p \qquad r = \frac{H}{6}$$

Fig 10.1 *ISO* metric thread form

Before measurement of a screw thread can be considered it will be necessary to identify the thread elements. These are shown at Figure 10.2 where the diagram represents a section of an external thread in an axial plane.

The elements of the thread shown at Figure 10.2 are as follows:

(*a*) *Major diameter.* (Sometimes called Full diameter.)

(*b*) *Minor diameter.* (Sometimes called Core diameter.)

(*c*) *Effective* (*pitch*) *diameter.* This is the diameter of the thread at which the thickness of the thread is equal to one half of the pitch $\left(\dfrac{p}{2}\right)$.

(*d*) *Pitch.* This is the distance, measured parallel to the screw axis, between corresponding points on adjacent thread contours.

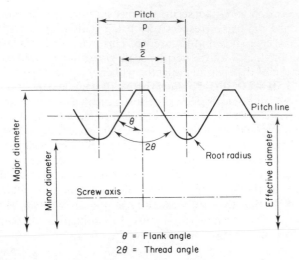

θ = Flank angle
2θ = Thread angle

Fig 10.2 External thread section

(*e*) *Thread angle.* This is the angle between the thread flanks measured in the axial plane (2θ). The flank angle is the semi-angle. (θ).

(*f*) *Root radius.*

As will be seen later, the most important of these elements are the effective diameter, pitch and flank angles as these most affect the fit of mating threads.

10.1 Measurement of thread gauge diameters

Sections 10.1, 10.2 and 10.3 respectively of this chapter are concerned with the practical activity of *measuring* all the *thread elements* of an external screw thread gauge, of the type shown at Figure 10.3. Here the TEC Unit is departed from slightly, in that the thread angle check is

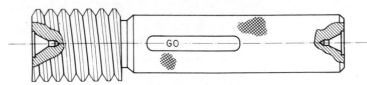

Fig 10.3 External screw thread gauge

included in the optical method outlined in Section 10.3. (See Learning Objectives 10 at the end of this chapter.) In fact, modern optical projectors are so accurate that for many applications they could be used to measure every one of the thread elements, but we will assume that in this case the measuring work on the thread diameters and pitch is to be carried out by the more common mechanical methods.

In this section we will describe the measurement of the major, minor and effective diameters respectively using a floating carriage micrometer of the type shown at Figure 9.5. This machine has a large micrometer drum which with the vernier scale gives direct readings of 0·0002 mm. The use of the fiducial indicator will enable the thread elements to be measured within 0·0025 mm irrespective of the operator's skill in determining micrometer 'feel'. The instrument is levelled by means of the adjustable screw in the base before measurement is started.

(*a*) *Measurement of major diameter.* The machine is first set to a datum (the word 'datum' means 'a given fact') by means of a plain cylindrical standard (plug gauge) of known diameter, the reading of the micrometer drum being noted. The standard is then replaced between the centres by the screw plug gauge, and a second micrometer reading taken. The set-up is shown diagrammatically at Figure 10.4.

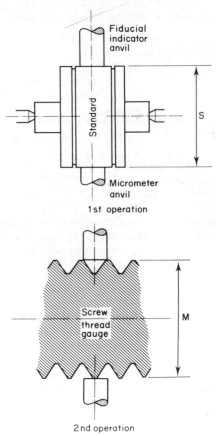

Fig 10.4 Measurement of major diameter

Referring to Figure 10.4:

Let   the micrometer reading over standard $= R_1$
               the micrometer reading over thread gauge $= R_2$
               the diameter of the standard $= S$
               the major diameter of the thread gauge $= M$

then $M = S \pm$ (difference between R_1 and R_2.)

The $+$ or $-$ is determined by whether the standard is a smaller or larger diameter than the thread gauge major diameter.

Example 10.1

When measuring the major diameter of an external screw thread gauge, a 30·500 mm diameter cylindrical standard was used. The micrometer readings over the standard and gauge were 9·3768 and 11·8768 mm respectively. Calculate the thread gauge major diameter (M).

Solution

$$R_1 = 9\cdot3768 \text{ mm}$$
$$R_2 = 11\cdot8768 \text{ mm}$$
$$S = 30\cdot500 \text{ mm}$$

Note that the standard is seen to be a nominally smaller diameter than the thread gauge full diameter:

$\therefore$
$$M = S + (R_2 - R_1)$$
$$= 30\cdot500 + (11\cdot8768 - 9\cdot3768)$$
$$= 30\cdot500 + 2\cdot500$$
$$= 33\cdot000 \text{ mm}$$

(b) *Measurement of minor diameter.* This is the same process in principle as that described for measuring the major diameter, but vee-shaped prisms are used of approximately 45° angle which sit in the roots of the threads. Prisms of suitable size are placed between the standard and the instrument anvils, and the first micrometer reading taken. The standard is replaced by the thread gauge, and a second micrometer reading is taken. The procedure is illustrated at Figure 10.5.

Referring to Figure 10.5:

Let   the micrometer reading over standard and prisms $= R_1$
      the micrometer reading over gauge and prisms $= R_2$
      the diameter of the standard $= S$
      the minor diameter of the thread gauge $= m$

then $m = S \pm$ (difference between R_1 and R_2)

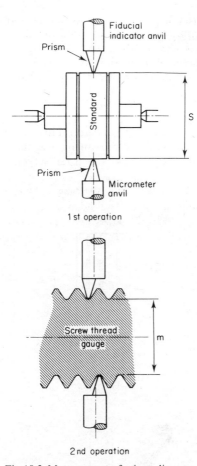

Fig 10.5 Measurement of minor diameter

Example 10.2

When measuring the minor diameter of an external screw thread gauge, a 30·500 mm diameter cylindrical standard was used. The micrometer readings over the standard and prisms, and gauge and prisms were 15·3768 and 13·5218 mm respectively. Calculate the thread gauge minor diameter (m).

Solution

$$R_1 = 15\cdot3768 \text{ mm}$$
$$R_2 = 13\cdot5218 \text{ mm}$$
$$S = 30\cdot500 \text{ mm}$$

Note that the standard is seen to be a nominally larger diameter than the thread gauge core diameter:

$$\therefore \quad m = S - (R_1 - R_2)$$
$$= 30 \cdot 500 - (15 \cdot 3768 - 13 \cdot 5218)$$
$$= 30 \cdot 500 - 1 \cdot 855$$
$$= 28 \cdot 645 \text{ mm}$$

(c) *Measurement of effective diameter.* This is the same process in principle as that described for measuring the major and minor diameters, but precision cylinders are used which lie in the thread

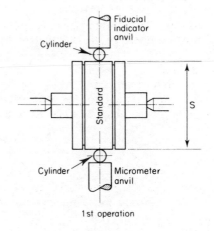

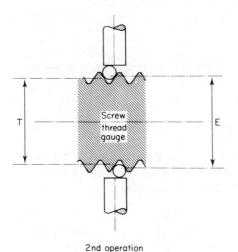

Fig 10.6 Measurement of effective diameter

grooves. Cylinders of suitable size are placed between the standard and the instrument anvils, and the first micrometer reading taken. The standard is replaced by the thread gauge, and a second micrometer reading is taken. The procedure is shown at Figure 10.6.

Referring to Figure 10.6:

Let   the micrometer reading over standard and cylinders $= R_1$

the micrometer reading over gauge and cylinders $= R_2$

the diameter of the standard $= S$

the diameter under the cylinders $= T$

the effective diameter of the thread gauge $= E$

then $E = T + P$

where P is a constant whose value depends upon the cylinder diameters and the thread angle. For an *ISO* metric thread:

$$P = 0 \cdot 86602 P - d$$

where $p =$ thread pitch, and

                $d =$ mean diameter of cylinders used.

Also $T = S \pm$ (difference between R_1 and R_2)

Example 10.3

When measuring the effective diameter of an external screw thread gauge of 3·5 mm pitch, a 30·500 mm diameter cylindrical standard and 2·000 mm diameter precision cylinders were used. The micrometer readings over the standard and cylinders, and gauge and cylinders were 13·3768 and 12·2428 mm respectively. Calculate the thread gauge effective diameter (E).

Solution

$$R_1 = 13 \cdot 3768$$
$$R_2 = 12 \cdot 2428 \text{ mm}$$
$$S = 30 \cdot 500 \text{ mm}$$
$$d = 2 \cdot 000 \text{ mm}$$
$$p = 3 \cdot 5$$

Note that the standard is seen to be a nominally larger diameter than the diameter under the cylinders:

$\therefore$
$$T = S - (R_1 - R_2)$$
$$= 30 \cdot 500 - (13 \cdot 3768 - 12 \cdot 2428)$$
$$= 30 \cdot 500 - 1 \cdot 1340$$
$$= 29 \cdot 3660$$

$$P = 0 \cdot 86602p - d$$
$$= (0 \cdot 86602 \times 3 \cdot 5) - 2 \cdot 000$$
$$= 3 \cdot 03107 - 2 \cdot 000$$
$$= 1 \cdot 03107 \text{ mm}$$

$$E = T + P$$
$$= 29 \cdot 3660 + 1 \cdot 03107$$
$$= 30 \cdot 39707 \text{ say } 30 \cdot 3971 \text{ mm}$$

When highly accurate measurement is being made of the effective diameter it would be necessary to make a correction for the thread rake and the compression which occurs due to the micrometer measuring force. In the above examples both correction factors are ignored.

10.2 Thread gauge pitch measurement

A relatively simple and accurate method of *measuring the pitch* of a screw thread is by means of a pitch measuring machine of the type shown at Figure 10.7.

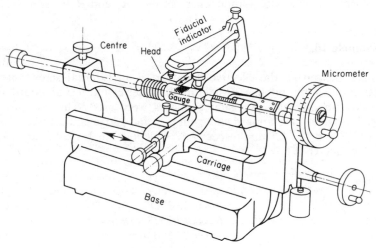

Fig 10.7 Pitch measuring machine

The screw gauge to be measured is mounted between centres, and the machine is fitted with a suitable stylus (the view of which is obscured by the thread gauge in Figure 10.7) of such a size, that it contacts the thread flanks approximately at the pitch line when it lies in the thread groove. This stylus is secured to a spring loaded head such that the stylus can ride up and down the thread crests as the carriage carrying the head is displaced along the bed parallel to the screw gauge axis. This is shown at Figure 10.8.

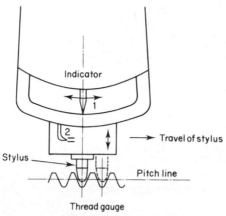

Fig 10.8 Pitch measurement

(*d*) *Measurement of pitch*. The principle of thread pitch measurement is that the stylus is moved along the thread, space by space, the amount of movement between each space being measured by the micrometer, this being used of course to displace the carriage along the bed. The pointer, number 1 (in Figure 10.8) is always brought opposite the index mark as each pitch reading is taken. When initially set up, the stylus is engaged in the thread groove with sufficient force such that pointer number 2 (in Figure 10.8) registers opposite the datum lines as shown. It is possible using this instrument to measure thread pitch to a similar accuracy as that for diameter measurement, i.e., 0·0025 mm.

10.3 Optical check of thread angle and form

Thread angles and *form* are usually checked by *optical* means, using an optical projector which may be of horizontal or vertical form. Cabinet projectors are available which are compact and take up relatively little space. Also measuring microscopes may be used, one form of which is the toolmaker's microscope which can also be used to carry out a complete check of the thread elements. Figure 10.9 shows the basic principle of optical projection, the general configuration being horizontal.

Figure 10.9 shows the four essential elements of a projection system, namely:

(i) *Source of light*. This is a lamp giving a point source, and is placed at the principal focus of the collimating lens.

(ii) *Collimating lens*. This gives a parallel light beam, which is essential in order to produce a sharp and accurate image.

(iii) *Projection lens*. This is a combination lens, and forms a real image on the screen of the object placed between it and the collimating lens.

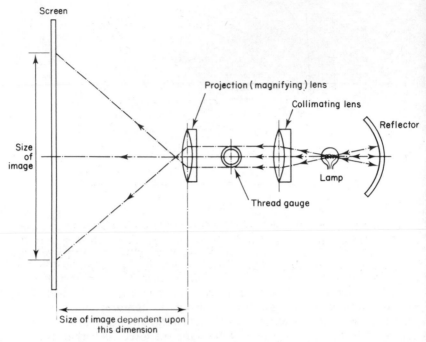

Fig 10.9 Principle of optical projection

(iv) *Screen.* This may be opaque or translucent.

For a clear definition of a thread form on the screen it is necessary to project the light beam along the line of the thread helix angle. This is done either by inclining the work through the appropriate helix angle, or by doing the same with the lamp and collimating lens unit. Both systems are in use on different makes of projectors.

(*e*) *Measurement of thread angle.* Most projectors are fitted with vernier protractors which may be read to 5 minutes of arc. Such a protractor is illustrated at Figure 10.10, the protractor being supported at the screen upon a straight-edge.

A vernier protractor, of the type shown at Figure 10.10 usually has a circular scale calibrated in half-degrees, minutes of arc being read off a tangent screw dial. Each flank angle is measured separately as shown, the protractor arm being rotated in turn to each flank of the projected thread image (shadow). Hence, the thread included angle (2θ) is the sum of the two flank angles (θ). (See Figure 10.2.)

(*f*) *Check on root radius.* Accurate templates of precise form and magnification are used with which to compare the projected shadow of the thread shape. These are available for optical projectors or microscopes. Figure 10.11 shows the view of a thread projected image and the

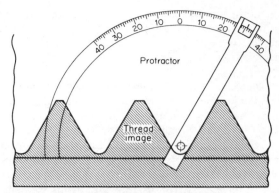

Fig 10.10 Vernier protractor

template as seen through the eyepiece of a toolmaker's microscope. The example shown is of a 2 mm pitch *ISO* metric thread. Any deviations of the root radius from the true form can clearly be seen by matching the thread magnified shadow with the template.

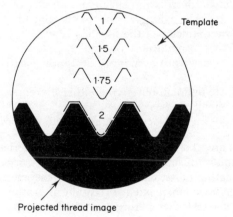

Fig 10.11 Projected thread image

10.4 Screw thread errors

Thread measurement, as previously described, may reveal *errors* in any one, or more, of the thread elements. However, errors in the effective diameter, pitch and thread angle are the most significant as these elements have the major effect on the fit of mating threads. Thread errors will be considered for each of the elements:

(*a*) *Major diameter*. Error in the major diameter may cause interference between mating threads or a reduction in the flank contact. Marginally, an incorrect major diameter might cause some component weakness, e.g., the major diameter of an external thread which is too small. Errors in major diameters of threads are generally caused by machine setting, e.g., the initial outside diameter to which the blank is machined controls the major diameter of an external screw thread.

(b) *Minor diameter*. Error in this element may similarly cause interference or flank contact reduction. Possible component weakness here could be caused by a reduction in root section, e.g., the minor diameter of an external thread being too small. Errors in minor diameters of threads are generally caused by machine setting, e.g., the initial hole diameter to which a blank is machined controls the minor diameter of an internal screw thread.

(c) *Effective diameter*. Error in this important element will cause either interference between the thread flanks or general slackness of fit, between mating parts. If the major and minor diameters are at the maximum limit, and the effective diameter below the minimum limit, then the thread will be thin on an external screw and thick on an internal screw. Errors in effective diameters of threads are generally caused by machine setting, e.g., the initial outside diameter to which a blank is machined controls the effective diameter of a rolled external screw thread. In this example, the blank outside diameter is the nominal effective diameter of the finished screw.

(d) *Pitch*. Errors in pitch may be:
  (i) periodic, varying in magnitude and recurring at regular intervals;
 (ii) progressive, uniform but giving either a greater or smaller pitch than nominal; or
(iii) erratic, varying and occurring at irregular intervals.

Pitch errors are likely to cause a progressive interference and tightening on assembly of mating parts. Such errors, which might be on one or both mating parts, can be related to the effective diameter. A pitch error can be compensated for by an alteration to the effective diameter, hence leading to mating parts being properly assembled together, i.e., the nut will screw on to the external thread.

Imagine a perfect nut apparently assembled with an otherwise perfect bolt having a progressive pitch error 'x' over a given length of thread engagement. This condition is shown at Figure 10.12(a), where θ is the flank angle.

The two threads clearly will not assemble in practice due to the interference. However, if the nut effective diameter is increased while the pitch remains unchanged then the threads will assemble without interference. A practical adjustment of this sort would clearly lead to a thinning of the thread flanks. Although the actual effective diameter of the bolt is the same as that of a bolt of perfect pitch, the Virtual

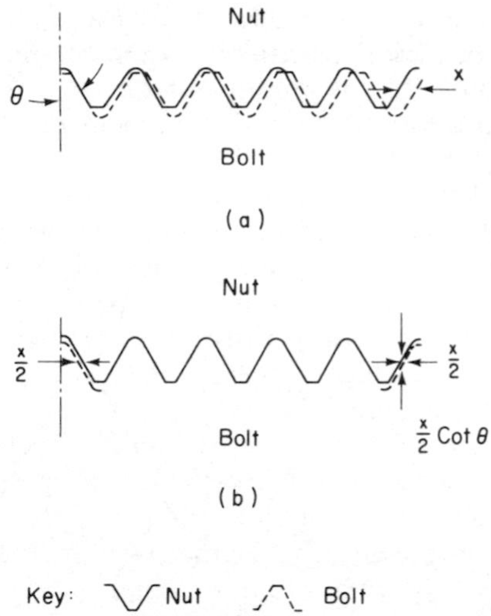

Fig 10.12 Pitch error

Effective Diameter is greater, since it is that of the now increased nut effective diameter.

Consider Figure 10.12(*b*) where the progressive pitch error 'x' is shown equally distributed at each end of the thread engagement. The vertical movement necessary to produce thread coincidence between the nut and bolt is $\frac{x}{2}$ Cot θ, therefore the diametral effect is x Cot θ. For an *ISO* metric thread $\theta = 30°$ so that x Cot $30° = 1.732x$, i.e., the value of any pitch error is virtually doubled when related to the effective diameter. Although this analysis assumed a progressive pitch error, it is equally true for periodic or erratic pitch errors.

Errors in thread pitch are generally caused during the cutting or finishing process, especially where a lead screw is the controlling factor which itself may have pitch errors. Where heat treatment is required, this may also cause pitch error due to distortion.

(*e*) *Thread angle.* Errors here can cause interference between mating threads, whether the flank angle is larger or smaller than the nominal angle. Consider Figure 10·13 where a perfect bolt is apparently assembled with an otherwise perfect nut having one incorrect flank angle, this error being δ radians.

The two threads clearly will not assemble in practice due to the interference. It can be shown that the vertical displacement necessary to give clearance is h δ Cosec 2θ, and this will be the diametral increase necessary on the effective diameter assuming δ is equally distributed between both thread flanks.

If the errors on both thread flanks are termed δ_1 and δ_2 respectively, measured in degrees, the diametral effect for an *ISO* metric thread of 60° thread angle is $0.0131 \times p(\delta_1 + \delta_2)$ where '*p*' is the thread pitch. Thread angle errors and pitch errors always increase the virtual effective diameter of an external thread.

Errors in thread angle may be caused by setting, but is a function of the accuracy of the thread form on the die, tool, etc., which is carrying out the thread production. In practice, such errors are rare.

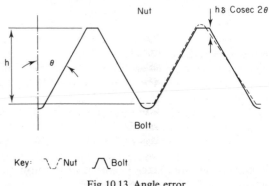

Fig 10.13 Angle error

(*f*) *Root radius.* An error at the root may cause excessive interference on the thread. Errors in the thread form here will be a function of the die, etc., being used as the medium for reproducing the thread shape.

10.5 Virtual effective diameter
This concept was raised in the preceding section under screw thread errors. The *virtual effective diameter* of a screw thread is the effective diameter of a nut of perfect pitch and form with which the screw will assemble with no diametral play. Similarly, the virtual effective diameter of a nut thread is the effective diameter of a screw of perfect pitch and form with which the nut will assemble with no diametral play.

The virtual effective diameter of an external screw is calculated by adding diametral equivalents of the pitch and flank angle errors of the screw to the screw effective diameter. For a nut, the virtual effective diameter is calculated by subtracting the diametral equivalents of the pitch and flank angle errors from its effective diameter.

Hence, for an external screw thread of *ISO* metric form:

$$
\begin{aligned}
\text{Virtual Effective Diameter} = \ & \text{Screw effective diameter} \\
& + 1.732x \\
& + 0.0131\, p\, (\delta_1 + \delta_2)
\end{aligned}
$$

[See Section 10.4(*d*) and (*e*).]

i.e., $VED = E + 1.732x + 0.0131\,p\,(\delta_1 + \delta_2)$

where E = screw effective diameter,
           x = total pitch error over a given
               engagement length,
           p = thread pitch,
           δ_1 = flank angle error in degrees (one side),
           δ_2 = flank angle error in degrees (opposite side).

Example 10.4

When measuring an *ISO* metric external screw thread gauge using the methods described in Sections 10.1, 2 and 3, the following results were recorded:

$$\begin{aligned}
\text{effective diameter} &= 30.6651 \text{ mm} \\
\text{pitch} &= 3.500 \text{ mm} \\
\text{pitch error} &= 0.004 \text{ mm} \\
\text{thread angle error} & \\
(RH \text{ flank}) &= 7 \text{ min. of arc} \\
\text{thread angle error} & \\
(LH \text{ flank}) &= 9 \text{ min. of arc}
\end{aligned}$$

Calculate the virtual effective diameter of the gauge.

Solution

$$\begin{aligned}
VED &= E + 1.732x + 0.0131\,p\,(\delta_1 + \delta_2) \\
&= 30.6651 + (1.732 \times 0.004) + [0.0131 \times 3.500(\tfrac{7}{60} + \tfrac{9}{60})] \\
&= 30.6651 + 0.0069 + 0.0122 \\
&= 30.6842 \text{ mm}
\end{aligned}$$

10.6 Taylor's principle applied to thread gauges

Although *Taylor's principle* is a topic in the TEC Unit U75/050 'Manufacturing Technology III', it will be worth while to re-examine it here before considering its application to the design of screw thread gauges.

    This important gauging principle, first formulated in 1905, states that the GO gauge should be of full form (i.e., it should check both geometric features and size), while the NOT GO gauge should check only one linear dimension. The truth of this principle can best be understood if one looks at an example of its application. Let us take the case of a rectangular hole, with a tolerance zone as shown in

Figure 10.14 which is to be gauged for both linear and geometric features.

Let us first see how the principle works if there is an error of geometry, say for example the corners of the rectangle are not square. Only a full form GO gauge will reject the work for this error. Simple

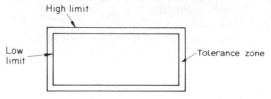

Fig 10.14 Rectangular hole

linear gauges of the pin gauge pattern, made to the hole low limits, will not detect this error, as it is possible that they will still enter the hole, indicating the hole is satisfactory. This condition is shown at Figure 10.15.

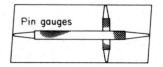

Fig 10.15 Hole with geometric error

Secondly, consider a linear error, or error of size; say for example the hole length is too great being outside the high limit. Only a NOT GO gauge which checks one linear dimension will reject the work for this error. A full form NOT GO gauge made to the hole high limits will not detect this error, as the gauge does not enter the hole because the width is within limits, although the length is outside limits. Again, such a gauge is indicating that the hole is satisfactory when it is not. This condition is shown at Figure 10.16.

Correctly designed gauges conforming to Taylor's principle will consist of a full form, rectangular GO gauge, and two NOT GO pin (linear) gauges for the hole width and length respectively. The pin

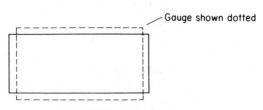

Fig 10.16 Hole with linear error

gauges are used to test the hole surface in several positions in order to detect possible errors. These gauges for a rectangular hole are shown at Figure 10.17.

The accuracy of modern thread production equipment is such that errors of thread form, including the thread angle, are very rare. Errors

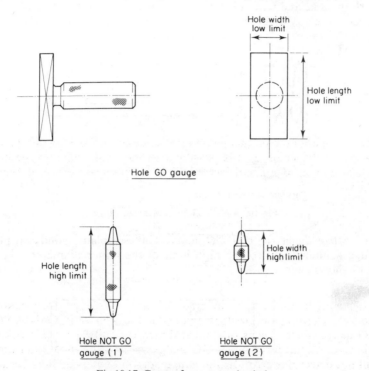

Fig 10.17 Gauges for rectangular hole

of the major and minor diameters are relatively unimportant, but the thread pitch and effective diameter must be carefully gauged during thread production as errors in these two elements vitally affect the fit of mating threads. Gauging of screw threads to Taylor's principle will be considered first for internal threads and secondly for external threads.

(i) Internal threads. The gauges required to satisfy Taylor's principle are:
    (*a*) Full-form GO gauge.
    (*b*) Effective diameter NOT GO gauge.
    (*c*) Minor diameter NOT GO gauge.
It is not usually necessary to gauge the major diameter. Each of these gauges will be briefly considered in turn.

(*a*) *Full form GO gauge.* This is an external screw plug gauge of the type illustrated at Figure 10.3, of a gauging length equal to the work length of engagement, and made to the low limits of size.

(*b*) *Effective diameter NOT GO gauge.* This screw plug gauge is made to the effective diameter high limit, havng restricted contact with the work threads, and with a length sufficient to accomodate two or three thread forms. A diagram of a typical gauge of this type is shown at Figure 10.18.

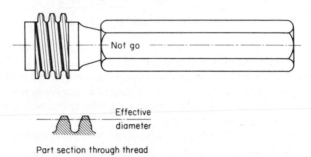

Part section through thread

Fig 10.18 NOT GO screw thread gauge

(*c*) *Minor diameter NOT GO gauge.* This is a plain cylindrical plug gauge manufactured to the high limit of the minor diameter. Figure 10.19 shows such a gauge.

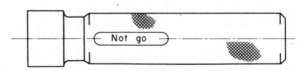

Fig 10.19 Plug gauge

(ii) External threads. In this case the gauges required to satisfy Taylor's principle are:

(*a*) Full-form GO gauge.

(*b*) Effective diameter NOT GO gauge.

(*c*) Major diameter NOT GO gauge.

It is not usually necessary to gauge the minor diameter. Again, each of these gauges will be briefly considered in turn.

(*a*) *Full form GO gauge.* This is usually a calliper-type gauge, having rollers or edge-type anvils, made to the high limits of size. Screw ring gauges are not now often used for this purpose as they are expensive to produce, and slow in use. A gauge of this type incorporates both the GO gauging anvils [required for (*a*)] and the NOT GO anvils [required for (*b*)].

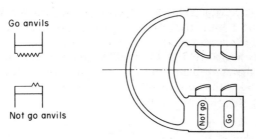

Fig. 10.20 Calliper screw thread gauge

(b) *Effective diameter NOT GO gauge.* This is a calliper gauge, made to the low limits of effective diameter and having restricted contact with the work threads in the same manner as illustrated in the part-section at Figure 10.18. The anvils for this gauge are incorporated with the full-form GO anvils, such a gauge being illustrated at Figure 10.20, where edge-type anvils are shown.

(c) *Major diameter NOT GO gauge.* This is a plain calliper, or gap gauge, made to the low limit of the thread major diameter. A gauge of this type is shown at Figure 10.21.

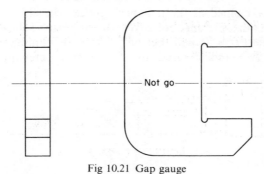

Fig 10.21 Gap gauge

10.7 Screw thread tolerances

Before examining *screw thread tolerances* it will be worthwhile to recapitulate the basic principles underlying a system of limits and fits. Figure 10.22 shows a mating shaft and hole having a clearance fit; of course a mating screw and nut must similarly have a clearance fit. Figure 10.22 shows the following factors:

(a) *Basic Size.* This is the nominal size specified by the designer, but which cannot be produced exactly because of the inherent inaccuracies of manufacturing processes.

(b) *Tolerance.* This is the amount of variation which can be tolerated to allow for the inherent inaccuracies of the manufacturing process. It is the difference between maximum limit and minimum limit.

(*c*) *Limits*. The maximum limit is the largest permissible dimension, and the minimum limit is the smallest permissible dimension.

(*d*) *Clearance*. The assembly shown has a clearance fit, the amount of clearance being affected by the chosen tolerance band.

One other important factor is the fundamental deviation (*FD*),

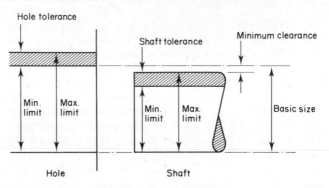

Fig 10.22 Hole and shaft limits

which can be defined as the distance between the basic size and the nearest end of the tolerance zone. With a basic size as shown at Figure 10.22 the hole $FD = 0$, and the shaft $FD =$ the minimum clearance.

The British Standard used for *ISO* metric screwthreads is BS

Class of fit	Tolerance class	
	Nuts	Bolts
Close	5H	4h
Medium	6H	6g
Free	7H	8g

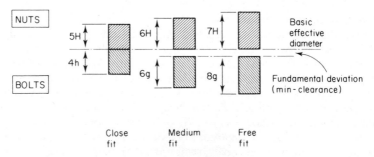

Fig 10.23 Classes of fit

3643: Part 2 and for any nominal (basic) thread diameter specifies (i) pitch, (ii) normal length of engagement, (iii) tolerance class, (iv) FD, (v) major diameter (min. limit), (vi) effective diameter (max. and min. limits), (viii) minor diameter (max. and min. limits).

In the *ISO* system, *FD's* are designated by capital letters for nuts and small letters for bolts. A number is used to designate the tolerance grade. The tolerance class [(iii) above] is a combination of the *FD* (letter) and tolerance grade (number), e.g., 5*H*. Although many tolerance classes are available within the *ISO* system giving a range of fits, most engineering requirements are provided for by three classes of fit, viz., close, medium and free. These are shown at Figure 10.23.

The letter '*H*' is used for nuts and the letter '*h*' is used for bolts where the *FD's* are zero. Reference to Figure 10.23 shows this. Reference to BS 3643: Part 2 shows that no tolerance is given for pitch. This is because of the relationship between pitch errors and the effective diameter [ref. Section 10.4 (*d*)]. It is therefore easier to control thread pitch errors by fixing a limit for their equivalent in terms of the effective diameter.

To fully specify an *ISO* metric screw thread it should be written as shown at Figure 10.24. The bolt thread specified at Figure 5.13 is taken as an example.

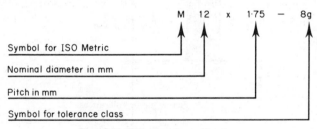

Fig 10.24 *ISO* thread specification

10.8 Screw gauge tolerances

The design and manufacture of gauges for screw threads is a highly skilled and specialist activity.

The British Standard Specification covering *screw gauge limits* and *tolerances* for *ISO* metric screw thread form is BS 919: Part 4. This specification includes GO and NOT GO screw plug, ring and calliper gauges and setting plugs. The specification is based on the usual gauging principle that the gauge-making tolerance zones are placed opposite to the direction of wear, i.e., inside the work limits for GO gauges, and outside the work limits for NOT GO gauges. This principle is illustrated at Figure 10.25.

Reference to Figure 10.25 shows a margin for wear on the GO size of a screw plug gauge. The gauge-making tolerances are approximately 10% of the work tolerance.

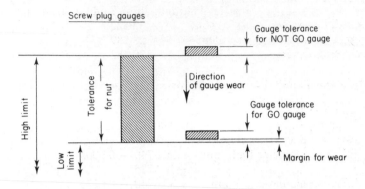

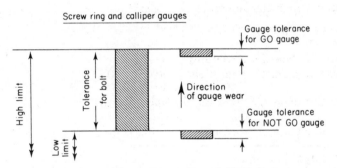

Fig 10.25 Screw gauge tolerances

An example will now be given to show the disposition of screw gauge tolerances:

Example 10.5

Consider an external screw thread of *ISO* metric form and coarse pitch, which is being produced in quantity. The thread has a nominal diameter of 33 mm, a pitch of 3·5 mm and a thread engagement length of 30 mm. (*a*) Show the limits for the thread diameters, and (*b*) show the limits for (i) a full-form GO screw calliper gauge, (ii) an effective diameter NOT GO screw calliper gauge, and (iii) a major diameter NOT GO plain calliper gauge.

Solution

(*a*) From BS 3643:Part 2 (mm units)

	High Limit	Low Limit	Tolerance
Major diameter	32·947	32·522	0·425
Minor diameter	28·653	28·189	0·464
Effective diameter	30·674	30·462	0·212

(*b*) From BS 919:Part 4 (mm units)

(i) GO gauge

Pitch tolerance	= 0·005
Effective diameter deviations	= −0·0005, −0·0235
∴ Effective diameter high limit	= 30·6740 − 0·0005
	= 30·6735
Effective diameter low limit	= 30·6740 − 0·0235
	= 30·6505
Effective diameter tolerance	= 0·023

(ii) NOT GO gauge

Effective diameter deviations	= 0, −0·023
∴ Effective diameter high limit	= 30·462 − 0
	= 30·4620
Effective diameter low limit	= 30·462 − 0·023
	= 30·4390
Effective diameter tolerance	= 0·023

(iii) NOT GO gauge

Gauge gap deviations	= 0, −0·0075
∴ Gap high limit	= 32·522 − 0
	= 32·5220
Gap low limit	= 32·5220 − 0·0075
	= 32·5145

10.9 Thread gauge applications

Vast numbers of both internal and external screw threads of varying accuracy are produced in the engineering industry each year. Depending upon the class of work, inspection can vary from a cursory glance to complete *gauging* or measurement of every thread element on every component part. For most general thread production it is not very usual to now find a full set of gauges in use as outlined in Section 10.6. Indeed, for reasons stated in that section, neither does one often see screw ring gauges and screw plug gauges of full form in use. The accuracy of modern dies and taps is such that only the effective diameter needs positive control.

Therefore, the most common way of gauging external threads is by use of a caliper screw gauge of the type illustrated at Figure 10.20. This type of gauge usually has adjustable anvils which are set the correct distance apart using setting plugs as described in BS 919. For internal threads, the tap is usually given a full inspection, after which inspection of the work using only a NOT GO effective diameter screw plug gauge of the type illustrated at Figure 10.18 is necessary.

Learning Objectives 10
The learning objectives of this section of the TEC unit are:

10. Understands the principles and appraises the applications of screw thread measurement and gauging.

10.1 Measures effective, major and minor diameters of an external screw thread gauge.

10.2 Measures, pitch and flank angle of an external screw thread gauge.

10.3 Uses optical methods to check form.

10.4 Lists, describes and relates to the manufacturing method, the common types of screw thread errors including those of pitch, form and angle.

10.5 Calculates virtual effective diameter.

10.6 Applies Taylor's principle to the design of screw thread gauges.

10.7 States the disposition of tolerances on screw threads with reference to the appropriate standards.

10.8 Uses 10.7 to establish gauge tolerances according to the appropriate standards.

10.9 Selects screw thread gauges for particular applications.

Exercises 10

1   Using a floating carriage micrometer and accessories, measure the major, minor and effective diameters of a screw plug gauge.

2   Using a pitch measuring machine, measure the pitch of a screw plug gauge.

3   Using an optical projector, or toolmaker's microscope, check the thread angle and form of a screw plug gauge.

4   Describe the common errors to be found in the several elements of a screw thread.

5   An external thread of *ISO* metric form has the following dimensions and errors:

effective diameter $= 21{\cdot}905$ mm

pitch $= 3{\cdot}000$ mm

pitch error $= 0{\cdot}0025$ mm

thread angle error (*RH* flank) $= 5$ min. of arc

thread angle error (*LH* flank) $= 6$ min. of arc

Calculate the virtual effective diameter of the thread. (Ans. 21·9165 mm)

6 Write down an expression for the virtual effective diameter of an internal thread of *ISO* metric form.

7 (*a*) State Taylor's principle of gauging.
(*b*) Show, using diagrams, how this principle may be applied to the design of internal and external screw thread gauges.

8 Obtain a copy of, and examine BS 3643, which covers limits and tolerances of screw threads.

9 Obtain a copy of, and examine BS 919 which covers limits and tolerances of screw thread gauges.

10 An *ISO* internal screw thread being produced in quantity has a nominal diameter of 24 mm, a pitch of 3·0 mm, and a thread engagement length of 35 mm. (*a*) Show the limits for the thread diameters, and (*b*) show the limits for (i) a full form GO screw plug gauge, (ii) an effective diameter NOT GO screw plug gauge, and (iii) a minor diameter NOT GO plain plug gauge.

Chapter 11
Measurement and Testing of Spur Gears

Before proceeding with the measurement and testing of spur gears it is necessary to understand the basic principles of the shape and structure of involute tooth gears. Reference to the introduction to Chapter 7, and Figures 7.1 and 7.2 will serve as an explanation of the principles. It is not usually necessary in practice to measure all the elements identified in Figure 7.2 once a blank of known diameter has had gear teeth generated upon it using an inherently accurate machine and cutter. Tooth thickness is commonly measured as this depends upon the depth to which the gear cutter is advanced into the blank.

However we will be concerned in this chapter with the measurement of the following important features:

(*a*) tooth thickness (see Section 11.2);
(*b*) tooth spacing (see Section 11.2);
(*c*) involute form (see Section 11.3);
(*d*) concentricity (see Section 11.3).

All of these features are important in regard to the accurate running of the gear in service.

11.1 Gear tooth elements
The following *gear tooth elements* will be identified here:

(i) chordal thickness;
(ii) constant chord;
(iii) base pitch;
(iv) base tangent.

(i) **Chordal thickness.** Reference to Figure 11.1 shows the chordal thickness to be the tooth thickness '*w*' at the reference circle, at a depth '*h*' from the top of the tooth. From Figure 11.1:

$$\frac{w}{2} = \frac{D}{2} \operatorname{Sin} \frac{90°}{T}$$

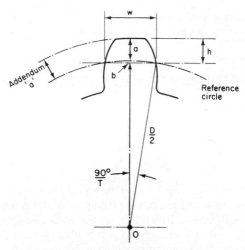

Fig 11.1 Gear tooth chordal thickness

$\therefore$
$$w = D \operatorname{Sin} \frac{90°}{T}$$

but
$$m = \frac{D}{T}$$

where D = diameter of reference circle; T = number of teeth in gear; m = gear module.

$\therefore$
$$w = mT \operatorname{Sin} \frac{90°}{T} \qquad (1)$$

$$h = \text{addendum} + \text{arc height}$$
$$= a + b$$
$$= m + \frac{D}{2} - \left(\frac{D}{2} \operatorname{Cos} \frac{90°}{T} \right)$$
$$= m + \frac{D}{2} \left(1 - \operatorname{Cos} \frac{90°}{T} \right)$$
$$= m + \frac{mT}{2} \left(1 - \operatorname{Cos} \frac{90°}{T} \right)$$

$\therefore$
$$h = m \left[1 + \frac{T}{2} \left(1 - \operatorname{Cos} \frac{90°}{T} \right) \right] \qquad (2)$$

The chordal thickness element is simple to understand, easy to measure and is correct for tooth forms other than involute. For these reasons it is widely used, but as can be seen from formulae (1) and (2) it depends upon the number of teeth in the gear.

(ii) Constant chord. Figure 11.2 shows an involute gear tooth symmetrically in mesh with a rack. The constant chord is seen to be the tooth thickness 'W' at the points of contact, at a depth 'H' from the top of the tooth.

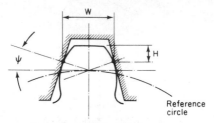

Fig 11.2 Gear tooth constant chord

The constant chord dimension (W) shown at Figure 11.2 is so called because it is fixed both by length and position, and is therefore independent of the number of teeth in the gear. It can be shown that:

$$W = \frac{m\pi}{2}\,(\cos^2 \psi) \tag{3}$$

$$H = m\left[1 - \frac{\pi}{4}\sin \psi \cos \psi\right] \tag{4}$$

where ψ = gear pressure angle.

The constant chord depends only upon the gear module and pressure angle, and is independent of the number of teeth in the gear as can be seen from formulae (3) and (4). It is not as easy to measure as the chordal thickness because it is smaller in magnitude and measured across a thinner portion of the tooth.

(iii) Base pitch. The base pitch (p_B) of a gear is the circular pitch of the teeth measured on the base circle. In Figure 11.3, the circular distance AB along the base circle is the base pitch.

Figure 11.3 shows the adjacent sides of two gear teeth having involute form flanks. From the properties of the involute, any tangential lines to the base circle drawn as shown, gives:

$$\text{base pitch } AB = CD \text{ etc.}$$
also
$$T \times AB = \text{base circle circumference}$$
$$= \pi D_B$$

where T = number of teeth in gear; D_B = diameter of base circle.

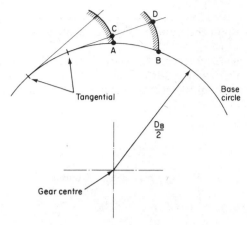

Fig 11.3 Gear tooth base pitch

$\therefore$
$$AB = \frac{\pi D_B}{T}$$

also
$$\frac{D_B}{2} = \frac{D}{2} \cos \psi \text{ (See Figure 7.2.)}$$

$\therefore$
$$D_B = D \cos \psi$$

where D = diameter of reference circle; ψ = pressure angle.

$\therefore$
$$AB = \frac{\pi D \cos \psi}{T}$$

but
$$m = \frac{D}{T}$$

$\therefore$
$$AB = m\pi \cos \psi$$

i.e., base pitch p_B $\qquad = m\pi \cos \psi \qquad (5)$

The base pitch is an important gear tooth parameter, but cannot easily be gauged as can the chordal dimensions described earlier. It can however be measured. The concept of the base pitch leads to that of the base tangent.

(iv) Base tangent. The geometry of the dimension known as the base tangent (t_B) is shown at Figure 11.4.

A base tangent measurement can be made across several teeth, and can be taken at various positions, such as *EF* or *GH* as shown at Figure 11.4. From the properties of the involute, any base tangents across opposed involute flanks are equal, i.e.,

base tangent *EF* = *GH* etc., = arc *JK*

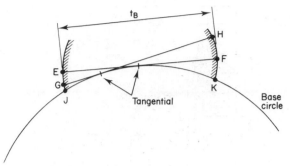

Fig 11.4 Gear tooth base tangent

It can be shown that:

$$t_B = p_B[0 \cdot 0047T + K - 0 \cdot 5] \text{ for } 20° \text{ pressure angle gears} \qquad (6)$$

where p_B = base pitch; T = number of teeth; K = number of teeth spanned by t_B.

The choice of the number of teeth (k) over which the base tangent is to be measured depends upon the number of teeth (T) in the gear. The following table gives a useful guide:

T	10–18	19–27	28–36	37–45	46–54	55–63	64–67	73–81	82–90
K	2	3	4	5	6	7	8	9	10

Commercial comparators are available for measuring base tangent, and tables are usually supplied from which the appropriate settings for t_B can be deduced. The base tangent is a most useful gear dimension as it can be taken across varying numbers of teeth in varying positions. This gives an indication of the accuracy of both the involute form and tooth spacing.

11.2 Measurement of gear tooth elements

(i) Measurement of chordal thickness. This is a measure of tooth thickness. The instrument most commonly used to *measure chordal thickness* is the gear vernier calliper shown at Figure 11.5.

The gear tooth vernier calliper is set, with the auxiliary slide at dimension '*h*', to measure the chordal thickness '*w*'.

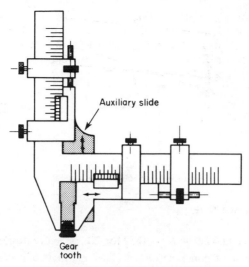

Fig 11.5 Gear tooth Vernier calliper

Example 11.1

Calculate the gear tooth calliper settings to measure the chordal thickness of a gear of 45 teeth having a module of 4.

Solution

From (1)

$$w = mT \, \mathrm{Sin} \, \frac{90°}{T}$$

$$= 4 \times 45 \, \mathrm{Sin} \, \frac{90°}{45}$$

$$= 180 \, \mathrm{Sin} \, 2°$$
$$= 180 \times 0·0349$$
∴ $$w = 6·28 \, \mathrm{mm}$$

From (2)

$$h = m\left[1 + \frac{T}{2}\left(1 - \mathrm{Cos} \, \frac{90°}{T}\right)\right]$$

$$= 4\left[1 + \frac{45}{2}\left(1 - \mathrm{Cos} \, \frac{90°}{45}\right)\right]$$

$$= 4[1 + 22·5 \, (1 - \mathrm{Cos} \, 2°)]$$
$$= [1 + 22·5 \, (1 - 0·9994)]$$
$$= 4[1 + 0·0135]$$
∴ $$h = 4·05 \, \mathrm{mm}$$

A simple gap gauge of the type illustrated at Figure 11.6 could be used for gauging chordal thickness, although this would of course require a degree of skill in deciding whether or not the gauge was a 'fit' on the tooth. For a given module, each gear having a different number of teeth would require a different gauge.

Gear
tooth

Fig 11.6 Gap gauge

(i) Measurement of constant chord. This again is a measure of tooth thickness. The instrument most commonly used to *measure* the *constant chord* is the gear tooth vernier calliper shown at Figure 11.5. The gear tooth calliper is set, with the auxiliary slide at dimension '*H*', to measure the constant chord '*W*'.

Example 11.2

Calculate the gear tooth calliper settings to measure the constant chord of a gear of 20° pressure angle having a module of 4.

Solution

From (3)

$$W = \frac{m\pi}{2} (\text{Cos}^2 \psi)$$

$$= \frac{4\pi}{2} (\text{Cos}^2 20°)$$

$$= 2\pi \times (0.9397)^2$$

∴ $W = 5.55$ mm

From (4)

$$H = m\left[1 - \frac{\pi}{4} \text{Sin} \psi \, \text{Cos} \psi \right]$$

$$= 4\left[1 - \frac{\pi}{4} \text{Sin } 20° \text{ Cos } 20°\right]$$

$$= 4\left[1 - \left(\frac{\pi}{4} \times 0\cdot3420 \times 0\cdot9397\right)\right]$$

$$= 4[1 - 0\cdot2524]$$

$$\therefore \qquad\qquad H = 2\cdot99 \text{ mm}$$

Again, a simple, solid gap gauge as shown at Figure 11.6 could be used to gauge the constant chord. The advantage here is that only one gauge is required for a given module.

(iii) Measurement of base pitch. This is a measure of tooth spacing. A convenient way to *measure base pitch* is by means of a height gauge, the method being shown at Figure 11.7.

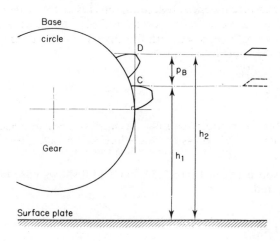

Fig 11.7 Measurement of base pitch

Comparison between Figures 11.3 and 11.7 will show that the difference in measured heights $(h_2 - h_1) = CD = $ base pitch.

Example 11.3

Calculate the base pitch for a gear of 20° pressure angle having a module of 4.

$$\therefore \qquad t_B = 11·83 \,[(0·0047 \times 45) + (5 - 0·5)]$$
$$= 11·83 \,[0·2115 + 4·5]$$
$$= 11·83 \times 4·7115$$
$$\therefore \qquad t_B = 55·74 \text{ mm}$$

i.e., the base tangent measured across 5 teeth = 55·74 mm.

Gear measurement using rollers. A simple additional check can be carried out using a pair of precision rollers of such a diameter that their centres lie on the reference circle of the gear. Measurement over the rollers placed in appropriate tooth spaces will give a measure of the tooth thickness and spacing. The geometry of the roller is best shown in relation to an involute rack tooth, as illustrated at Figure 11.9.

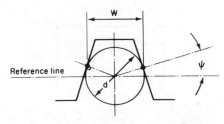

Fig 11.9 Roller and rack tooth geometry

Reference to Figure 11.9 will show that the roller must contact the rack tooth at the constant chord (W) already described at Figure 11.2.

Hence

$$\frac{\dfrac{W}{2}}{\dfrac{d}{2}} = \text{Cos } \psi$$

where d = roller diameter.

$$\therefore \qquad \frac{d}{2} = \frac{\dfrac{W}{2}}{\text{Cos } \psi}$$

but

$$W = \frac{m\pi}{2}\,(\text{Cos}^2\,\psi) \qquad (3)$$

$$\therefore \qquad \frac{d}{2} = m\pi\,\frac{(\text{Cos}^2\,\psi)}{4\,\text{Cos }\psi}$$

$$\therefore \qquad d = \frac{m\pi}{2}\,\text{Cos }\psi \qquad (4)$$

Note that the roller size is independent of the number of teeth in the gear. Measurement over the rollers is best carried out using a vernier calliper, although great care is needed to obtain accurate results.

Example 11.5

Calculate (*a*) roller diameter, (*b*) distance over rollers in the nearest to directly opposite, tooth spaces, (*c*) distance over rollers spaced 8 teeth apart, for a gear of 20° pressure angle, having 45 teeth and a module of 4.

Solution

(*a*) From (7)

$$d = \frac{m\pi}{2} \cos \psi$$

$$= \frac{4\pi}{2} \cos 20°$$

$$= 2\pi \times 0{\cdot}9397$$

∴ $$d = 5{\cdot}90 \text{ mm}$$

(*b*) As the gear has an odd number of teeth, the two rollers cannot lie in directly opposite tooth spaces, hence the geometry is as shown at Figure 11.10.

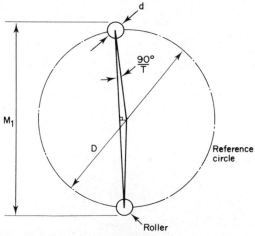

Fig 11.10 Measurement over rollers

From Figure 11.10

$$M_1 = D \cos\left(\frac{90°}{T}\right) + d$$

where M_1 = distance over rollers; D = diameter of reference circle; T = number of teeth; d = diameter of rollers.

Now
$$\begin{aligned} D &= mT = 4 \times 45 \\ &= 180 \text{ mm} \end{aligned}$$

Now
$$\begin{aligned} M_1 &= D \cos\left(\frac{90°}{T}\right) + d \\ &= \left[180 \times \cos\left(\frac{90°}{45}\right)\right] + 5·90 \\ &= [180 \times \cos 2°] + 5·90 \\ &= [180 \times 0·9994] + 5·90 \\ &= 179·892 + 5·90 \end{aligned}$$
$$\therefore \quad M_1 = 185·80 \text{ mm}$$

(c) The angle subtended by 8 teeth

$$= 360° \times \frac{8}{45} = 64°$$

The geometry is as shown at Figure 11.11.

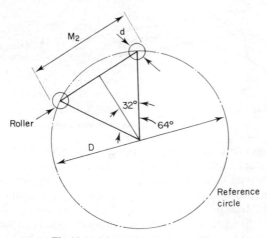

Fig 11.11 Measurement over rollers

$$M_2 = D\mathrm{Sin}\ 32° + d$$

where M_2 = distance over rollers; D = diameter of reference circle; d = diameter of rollers.

$$\therefore \quad M_2 = [180 × \mathrm{Sin}\ 32°] + 5·90$$
$$= [180 × 0·5299] + 5·90$$
$$= 95·382 + 5·90$$
$$\therefore \quad M_2 = 101·28\ \mathrm{mm}$$

The same treatment at (c) above could of course be applied to the situation at (b) in which case the solution would be:

$$M_1 = [180 × \mathrm{Sin}\ 88°] + 5·90$$
$$= 179·892 + 5·90$$
$$= 185·80\ \mathrm{mm}$$

11.3 Gear testing machines
This section will be concerned with the *testing* of the *involute form* and *concentricity* of gears, and the machines which are used for this purpose.

Involute testing machine. Figure 7.1 showed an involute curve as that described by the end of a tightly wrapped cord unwound from a base cylinder (circle). A similar, but alternative definition is that an involute is a curve swept out by the end of a straight edge rolled on a base circle. This is shown at Figure 11.12.

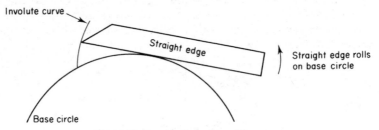

Fig 11.12 Construction of involute curve

The arrangement shown is the geometric basis of involute testing machines, except that the base circle now rolls along a straight edge. Generally on such machines, the gear to be tested is mounted, integral with a base disc (circle) of correct diameter, on a spindle which is housed in a longitudinal slideway. The disc is then rolled along a straight edge fixed to a cross-slide. An indicator is mounted immediately above the straight-edge, and has a stylus which is placed in contact with the gear tooth involute flank which is to be tested. The point of this fixed stylus is coincident with the edge of the straight-edge which is tangential to the base circle. The set-up is as shown at Figure 11.13.

If the involute curve is correct, then no deviations will be recorded

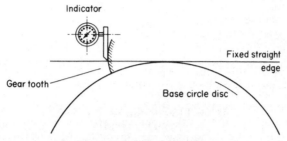

Fig 11.13 Testing of involute

upon the indicator. Commercial involute testing machines usually include a rectilinear recorder which produces a graph of the magnified movements of the indicator stylus in contact with the involute gear tooth. If the resulting graph is a straight line then the involute is correct. Any inherent errors are shown upon the graph as deviations from a straight line.

Rolling gear testing machine. The general arrangement of machines of this type is such that the gear to be tested for concentricity is mounted in mesh with a hardened and ground 'master' gear which is free from error. The two gears are slowly rotated together, and any errors due to eccentric running of the gear under test are shown upon an indicator, or again upon a recorder graph. The basic elements of such a machine are shown at Figure 11.14.

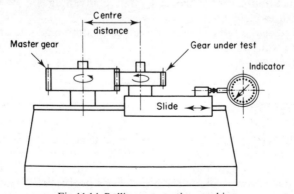

Fig 11.14 Rolling gear testing machine

The slide of the machine shown at Figure 11.14 is spring loaded. Hence, a rolling test is given under this spring loading with the gears set, using a vernier or slip-gauges, at the exact centre distance. Although such machines are primarily intended to indicate eccentric running, with experience other errors may be identified.

11.4 Gear tooth errors

Before the gear tooth spaces are machined out by a generation process, it is necessary to produce an accurate blank. The finished gear and its performance is influenced by the condition of the blank, particularly with regard to concentricity. For an accurate *gear* with minimum *error*, the blank must be produced to a high degree of accuracy of dimension, concentricity, roundness and squareness of the gear faces.

The accuracy of the tooth form and spacing is a function of the accuracy of the generating machine and cutter. Modern machines and cutters are made to such standards of accuracy, that a high degree of accuracy in the finished gear is possible if required. Errors in respect to tooth form and spacing are therefore rare. The one gear element which can most be affected during manufacture is tooth thickness, as this is a function of the depth setting of the cutter. Also cutter wear can effect this element.

Learning Objectives 11

The learning objectives of this section of the TEC unit are:

11. Understands the principles and appraises the applications of gear measurement and testing related to spur gears.

11.1 Identifies, given a spur gear or drawing of a gear, the following elements to be measured: Chordal thickness, base pitch, base tangent and summarises their relative advantages.

11.2 Measures elements listed in 11.1.

11.3 Describes the use of the involute gear testing machine and rolling gear testing machine.

11.4 Relates gear errors to manufacturing methods and performance.

Exercises 11

1   Identify the following gear tooth elements giving the relative merits and disadvantages: (*a*) Chordal thickness, (*b*) constant chord, (*c*) base pitch, (*d*) base tangent.

2   Derive mathematical expressions for the gear tooth elements listed in Question 1 above.

3 Calculate the gear tooth vernier calliper settings to measure (a) the chordal thickness, and (b) the constant chord, of a gear of 20° pressure angle, having a module of 3 and 24 teeth. (Ans. (a) $w = 4.71$ mm, $h = 3.08$ mm, (b) $W = 4.16$ mm, $H = 2.24$ mm.)

4 Calculate (a) the base pitch, and (b) the base tangent, of a gear of 20° pressure angle, having a module of 3 and 24 teeth. (Ans. (a) $p_B = 8.86$ mm, (b) $t_B = 23.15$ mm.)

5 Calculate (a) the roller diameter, (b) distance over rollers in directly opposite tooth spaces, (c) distance over rollers spaced 6 teeth apart, for a gear of 20° pressure angle, havng a module of 3 and 24 teeth. (Ans. (a) $d = 4.43$ mm, (b) $M_1 = 76.43$ mm, (c) $M_2 = 55.34$ mm.)

6 Sketch and describe the principle of operation of an involute gear testing machine.

7 With the aid of sketches, describe the principle of operation of a rolling gear testing machine.

8 Describe the common errors to be found in manufactured gears relating to the method of manufacture and gear performance.

Chapter 12
Geometrical Tolerancing

It is often necessary to specify and control the geometric features of a component, such as straightness, flatness, etc., as well as linear dimensions. Geometric tolerances are concerned with the accuracy of the relationship of one feature to another, and it is accepted that these should be separately specified.

BS 308 : Part 3 is a comprehensive specification of geometrical tolerancing, and gives the following definitions:

(i) *Geometrical tolerance*: The maximum permissible overall variation of form, or position of form, or position, of a feature.

(ii) *Tolerance zone*: The zone within which the feature is to be contained. This zone may be a circle, cylinder, area between two parallel lines, etc. Many tables are provded in the standard to illustrate tolerance zones.

On drawings, the geometrical tolerance is indicated in a rectangular frame, which is divided into compartments, as shown at Figure 12.1.

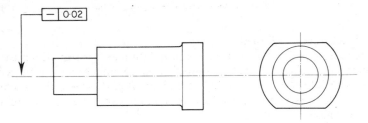

Fig 12.1 Geometrical tolerance frame

The geometrical tolerance is indicated separately from linear tolerances, using a frame illustrated in Figure 12.1 which has two compartments, although in some circumstances there may be more. The symbol for the characteristic being toleranced is placed in the left-hand compartment, the symbol at Figure 12.1 meaning 'straightness'. BS 308:Part 3 contains a table of tolerance symbols, some of which are shown at Figure 12.2. The tolerance value, in the units used for

linear dimensions, is placed in the next compartment, this being 0·02 mm at Figure 12.1. Hence, the geometrical tolerance indicated at Figure 12.1, means that the axis of the whole part is required to lie between two parallel straight lines 0·02 mm apart, in the plane of projection shown.

Characteristic to be toleranced	Symbol
Straightness	—
Flatness	⟋▱
Roundness	○
Cylindricity	⌭
Profile of a line	⌒
Profile of a surface	⌓
Parallelism	//
Squareness	⊥
Angularity	∠
Position	⊕
Concentricity	◎

Symmetry.

Fig 12.2 Geometrical tolerance symbols

12.1 Tolerance 'build-up'

Tolerance build-up, or tolerance accumulation as it is sometimes called, refers to the cumulative effect of individual tolerances on length, centre distances of holes, or assembly of individual parts where the assembly overall length is the sum of the individual part lengths. Figure 12.3 shows these conditions.

From Figure 12.3 it can be seen that:

overall basic length = sum of individual basic lengths.

i.e., $$L = X + Y + Z$$

and cumulative tolerance = sum of individual tolerances

i.e., $$p + q = (a + b) + (c + d) + (e + f)$$

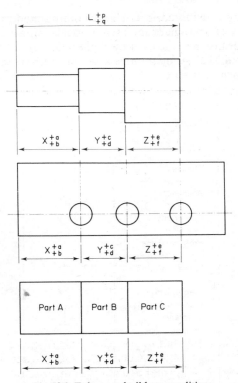

Fig 12.3 Tolerance build-up conditions

It can be seen from Figure 12.4 for example that:
overall basic length $= 50 + 72 + 36 = 158$
and cumulative tolerance $= (3 + 2) + (3 + 2) + (3 + 2) = 15$

In practice, the cumulative effects of the individual tolerances leads to an unacceptably high tolerance on the overall length, and it be-

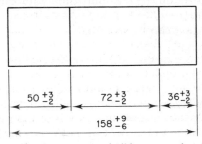

Fig 12.4 Tolerance build-up example

comes necessary to minimise this effect. One method of reducing the accumulation is to arrange progressive dimensioning from a common datum, one example being shown at Figure 12.5.

Two other points are worth consideration with regard to tolerance build-up on assembled parts. The first is that a final machining oper-ation could be carried out to finish the overall assembly length to the

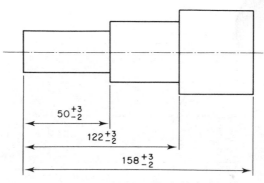

Fig 12.5 Progressive dimensioning

required limits, hence avoiding the effects of tolerance build-up. The second is that when a random choice is made from different com-ponent parts made to a normal distribution of sizes, the chance of two or more parts being at the extreme limits of size is very remote. The most common occurrence is that sizes around the average will result from random selection, extreme sizes being rarely selected. Hence, in such cases the adverse effects of tolerance accumulation are the exception.

12.2 Maximum and minimum metal conditions

A *minimum metal condition* is that where a part has the minimum amount of metal permitted by the dimensional tolerance, e.g., low limit for a shaft and a high limit for a hole. Conversely, a *maximum metal condition* (MMC) is that where a part has the maximum amount of metal permitted by the dimensional tolerance, e.g., high limit for a shaft and low limit for a hole. MMC is indicated on a drawing by the symbol M which is included in the tolerance frame. (See Figure 12.7.) MMC has a special importance with regard to geometrical tolerancing as it critically affects the interchangeability of manufactured parts which are to be assembled together. This is discussed in Section 12.3.

Figure 12.6 shows a component at both the minimum and maxi-mum metal condition.

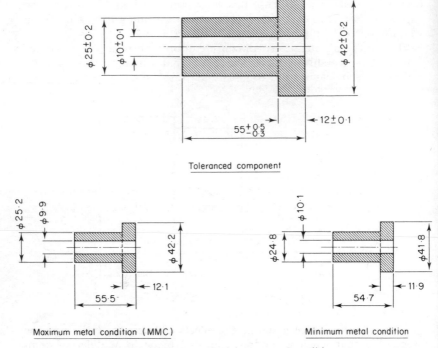

Fig 12.6 Maximum and minimum metal conditions

12.3 Virtual size

Where two components are required to freely assemble together, as shown at Figure 12.7, the assembly is dependent upon the combined effect of the linear and geometric errors. When a part has errors of form, or position, its size is virtually altered, i.e., the *'virtual' size* of the shaft is larger than specified. The opposite is true of the hole.

At Figure 12.7 the shaft is shown dimensioned, the length being ignored for the purposes of this example. The tolerance frame indicates that a geometrical straightness tolerance of 0·03 mm has been allowed, i.e., the shaft axis is required to be contained within a cylindrical zone of 0·03 mm diameter. Also, a diametral linear tolerance of 0·05 mm is specified. The worst condition for assembly of the shaft and hole occurs when the mating features are in the maximum metal condition (*MMC*), and in addition the maximum errors permitted by the geometric tolerance are present. This condition is shown at Figure 12.8.

In the worst condition, shown at Figure 12.8, the shaft will just be able to enter a truly cylindrical hole of diameter equal to the high limit shaft diameter (*MMC*) plus the geometrical tolerance, i.e., 30·56 mm diameter. This is now its virtual size. If all the linear tolerance is not

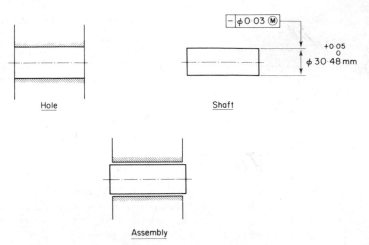

Fig 12.7 Shaft and hole assembly

taken up, and the shaft for example is at the minimum metal condition, i.e., 30·48 mm, then a greater geometrical tolerance than that specified is available without endangering free assembly. In this example, the geometrical tolerance could therefore be increased to 0·08 mm.

In all practical design cases it should be established whether or not an increase in the specified geometrical tolerance associated with the

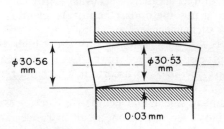

Fig 12.8 Worst assembly condition

MMC principle is acceptable. Where such an increase in geometrical tolerance can be allowed, the symbol *M* should be included in the tolerance frame, as shown at Figure 12.7. This is in accordance with the recommendations outlined in BS 380 : Part 3. The measurement of the linear and geometrical features discussed in this section is shown at Figure 12.9.

12.4 Geometrical tolerancing problems
It is suggested that *geometrical tolerances* should be specified for all features of parts critical to functioning and interchangeability, the ob-

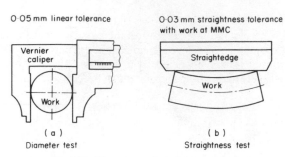

0·05 mm linear tolerance   0·03 mm straightness tolerance
with work at MMC

(a)   (b)
Diameter test   Straightness test

Fig 12.9 Diameter and straightness check

vious exception being when it is certain that the production techniques being used can be relied upon to achieve the required standard of accuracy. Geometrical tolerances, however, are not always found upon part drawings, even when geometric accuracy is important, because it is implied that an allowance for geometric error is contained within the linear tolerance. The main problem of geometric tolerancing is that it must be measured (as shown at Figure 12.9), in addition to the linear check. This, of course, increases the cost of the product, and a balance must be sought between the importance of maintaining the accuracy of a particular geometric feature and the extra costs incurred. Further, inspection judgements become more difficult when the implications of the *MMC* principle, discussed in the last section, must be taken into account. If the required geometric accuracy can easily be achieved by normal manufacturing methods, then it is clearly uneconomic to specify and check such a degree of accuracy. For example, a good quality lathe can be expected to consistently produce a high degree of round-ness and straightness on turned parts. Only an occasional random check is necessary in order to ascertain whether or not accuracy is deteriorating.

The relationship of linear and geometric tolerances is described in the last section. However, one geometrical tolerance can also affect an-other. For example, if a parallelism tolerance is applied to two opposite plane surfaces, then this will automatically limit the errors of flatness of surfaces. This is because the tolerance zone for flatness is the space between two parallel planes, which in turn may be the tolerance zone for parallelism. Each application of geometrical tolerances depends upon the functional requirements, interchangeability and the manu-facturing method of the part concerned. BS 308:Part 3 offers many typical applications for a variety of situations.

12.5 Economic aspects

Manufacturing costs are closely linked to the specified tolerances, of whatever type. The relationship is shown graphically at Figure 12.10.

The graph at Figure 12.10 shows that cost decreases as tolerance increases. Or, to put it another way, an increase in accuracy leads to an increase in cost. It is therefore obvious that finer tolerances than are necessary should not be specified by a designer. In deciding upon the appropriate tolerance, three factors must be considered, these being

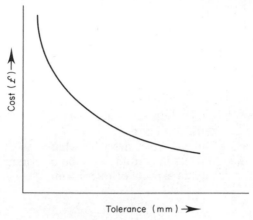

Fig 12.10 Tolerance-cost chart

the functional requirements, interchangeability and cost respectively. Interchangeability means that any standardised parts produced can be interchanged such that any component will assemble equally well with any mating part without any fitting being necessary. Therefore, this will only be possible when the class of fit and the appropriate tolerances have been decided upon, and of course cost is an important and related element.

In a highly accurate class of work, the high cost of maintaining the required degree of accuracy may be uneconomic. In such circumstances a selective system of assembly is used, such that components of the same nominal size are graded into batches according to the variation in size from the nominal. Therefore, the components in one batch are virtually identical in size. A good example of this technique will be found in the field of ball or roller bearing manufacture. Here, machines automatically select and grade the balls or rollers into batches just prior to assembly. A selective system of assembly prevents complete interchangeability, but allows coarser tolerances to be used during manufacture, hence reducing costs.

Learning Objectives 12
The learning objectives of this section of the TEC unit are:

12. Understands geometrical tolerancing.

12.1 Determines effect of tolerance 'build-up'.
12.2 Determines maximum and minimum metal conditions for a given component.
12.3 Calculates virtual size with reference to 12.2.
12.4 Explains problems associated with geometrical tolerancing.
12.5 Identifies the economic aspects of tolerancing.

Exercises 12

1 Using examples of your own choice, describe the effect of tolerance build-up.
2 Explain what is meant by, the terms 'maximum metal condition', and 'minimum metal condition', and show the relationship between *MMC* and geometrical tolerances.
3 With the aid of an example, describe what is meant by 'virtual size', and show how its magnitude may be determined.
4 Describe the economic aspects of tolerancing.

Chapter 13
Control Charts

This chapter is concerned with the control of quality upon the shop floor. There is now an increasing use of statistical techniques for this purpose, so much so, that the term quality control is often taken to mean the control of quality by statistical means. The theory of statistics and probability are brought together for this purpose.

Statistical sampling techniques require the inspector upon the shop floor to take controlled random samples of components which will then be checked by measurement or gauging. The quality of the whole batch of work produced will then be judged by the results of the sample, i.e., if the number of rejects found in the sample is too high then the quality of the whole batch is unacceptable; and vice versa. It is rather like taking a bite out of a large cake, and judging the quality of the whole cake upon the evidence obtained from tasting one mouthful, rather than eating the whole cake in order to be certain. This is where the idea of probability comes in. There is obviously a probability that the quality of the one mouthful of cake is different from the total quality level of the whole cake; either better or worse. The statistical part of the exercise is in collecting the data from the samples, in order that the total quality level can be assessed within a certain degree of probability. Hence, the inspector is as much concerned with numbers, charts, data, etc., as he is with measuring instruments and gauges.

The alternative to the above is to carry out 100% inspection such that every component leavng the production line is checked. This would appear to be more expensive, although the implication is that the results would have more certainty. However, this has been proved to be wrong, and there is evidence to show that 100% inspection does not guarantee that all unacceptable work will be found and rejected.

This is particularly so in the case of repetitive and boring activities, and let us face it, much engineering manufacture work is in this category. In certain classes of work, where the part must be destroyed in order to check its quality, statistical sampling techniques must be used.

13.1 Justifying statistical control

Inspection of work by skilled inspectors is expensive and time-consuming work. Gauging of work by operators and/or inspectors is monotonous and unreliable. The modern alternatives, which are more efficient, are: (i) automatic inspection by machines. This is known as 'auto-sizing' or 'in-process control'; (ii) quality control using *statistical* methods. The latter method has the great advantage of being the much *cheaper* method.

The use of control charts enables the sample results of patrol inspection taken at the machine to be used to analyse whether or not production is proceeding satisfactorily, and if corrective action needs to be taken. Hence, the main purpose of the chart is to indicate changes in quality so that adjustments can quickly be made to correct the process before large quantities of scrap are produced, i.e., in effect the chart gives advance warning of the commencement of a trend towards the production of an increasing number of defective articles. Control charts help to reduce the number of times that an adjustment is made to a machine set-up because the operator and inspector have a visual picture of the accuracy of the process. This, therefore, helps to reduce variability. Trends which become clear through the use of control charts help to determine the optimum time for tool changes because of tool wear.

In addition to the main purpose of control charts described above, they can be useful for:

(a) showing the inherent capabilities of a process for producing components of a particular accuracy, i.e., they give a profile of the inherent accuracy of a machine or process;

(b) recording the degree of accuracy of incoming materials which are subject to 'acceptance sampling';

(c) allowing a feed-back of information which can be used for planning and control purposes. This information can be used to help improve product quality and reduce production costs.

In certain circumstances, such as where a product has to be tested to destruction in order to prove that its quality is within the required specification, the use of sampling techniques is the only feasible way in which quality can be controlled. Much ordnance work, for example, fits into this category. It is often thought that control charts are only suitable for high quantity production processes where changes to the product are rare. It has however been proved that control charts are useful for short-run production processes, particularly in regard to the information which they give to design and planning staff.

To sum up, the most important characteristic of control charts is to give accurate information of the variability of a manufacturing process, and they help to identify assignable causes of variation as opposed to inherent process variation. (See Section 13.2)

13.2 Manufacturing variability

No two parts can be produced with identical measurements by a *manufacturing* process, and some *variability* of size within certain limits must be tolerated during manufacture. No matter how precise the process, it is subject to small fluctuations from a variety of causes which combine to cause variations in the product. This inherent process variation is a characteristic of the process and is the result of random causes. These random fluctuations cause the process to deviate either side of the average objectives. Some common random causes of variability in machined metal products for example, are:

(i) variations in the properties of the material;
(ii) variations in the type and supply of coolant;
(iii) temperature fluctuations;
(iv) condition of machine. A worn machine having excessive clearance in bearings and slides can cause vibrations during cutting, eccentric running of bar or cutter, etc.

In addition to random causes of variability described above, there are other sources of variability which cause a process to deviate further from its average than otherwise would be the case with only random causes. These are more obvious causes of variability known as assignable causes; i.e., it is possible to precisely assign a cause to a particular type of variation. Two common assignable causes of variability in machined metal products are:

(i) tool wear;
(ii) errors of tool setting.

Tool wear is the classic case of a cause of variability which can clearly be identified upon a control chart which records average values, as it causes a drift of the average size of samples away from that specified.

13.3 Control charts for variables

The appropriate British Standard to consult for the construction of *control charts* is BS 2564 Control Chart Technique (when manufacturing to a specification with special reference to articles machined to dimensional tolerances). Before proceeding with the construction of control charts, it would be appropriate to recall some basic principles of statistical sampling:

If samples of 'n' parts of size 'x' are taken at random from a controlled process which gives approximately a normal distribution, the means of the samples will themselves be normally distributed having a smaller amount of spread or scatter. Note that any distribution of means ($\bar{x}$) will tend to be more normal in shape than the original distribution of individual (x) values. Again note that the spread of the original distribution of individual values is defined by the range (w), or standard deviation (s).

The average value of a distribution of means is $\bar{X}$ (grand mean), and the standard deviation is $s_n = \dfrac{s}{\sqrt{n}}$. Similarly, the standard deviations or ranges of distributions of individual sizes, could themselves be distributed. The distribution of ranges is important, because it is the one used for control charts, 'w' being easier to calculate than 's'. This distribution will be found to be skew, i.e., the distribution of results is not symmetrically disposed around the mean ($\bar{w}$), the spread on one side of the mean being greater than the spread on the other side.

Further, with a controlled production process virtually all the parts produced fall by size within a distribution, the spread (variability) of which is equal to a band of 6 standard deviations width. With control charts therefore, it is customary to position the outside (action) limits 6 standard deviations apart. If inner (warning) limits are used, these are in turn positioned 4 standard deviations apart.

A *control chart for variables* is used for a manufactured part which the inspector checks by measurement, not gauging. This is less common, and of course this method of control is more expensive. However, it reveals much more about the behaviour of the process. With sample data in the form of measurements, it is possible to use a control chart

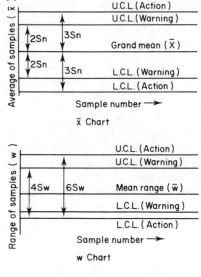

S_n = standard deviation of distribution of means
S_w = standard deviation of distribution of ranges
U.C.L = upper control limit
L.C.L = lower control limit

Fig 13.1 $\bar{x}$ and 'w' charts

showing deviation in the process mean ($\bar{x}$ chart), and one showing deviation in variability (w chart). The procedure for compiling these charts is that data has to be collected initially to establish the value of $\bar{X}$ (grand mean) and $\bar{w}$ (mean range) from a series of samples (usually 5 or 10 in number), and control limits can then be calculated. The charts will appear as shown at Figure 13.1.

Once the charts have been drawn up complete with the limits in the correct positions, as shown at Figure 13.1, then the results of each random sample of component sizes are plotted, using points, upon the control charts. Points thus plotted on the charts will indicate any tendency for the process to go out of statistical control. Any change of central tendency shown on the $\bar{x}$ chart may be due to tool wear, temperature increase, new work materials, etc. Any change of variability shown on the 'w' chart is more difficult to account for, but may be due to wear in machine bearings or slides, affect of careless operator (where one is employed), etc. The L.C.L.'s on this chart are less important than the U.C.L.'s, because a trend towards the lower limits indicates an improvement of quality as the variability is less.

With $\bar{x}$ and 'w' charts, the position of the limits is not calculated from first principles because the computation of the standard deviations can be very tedious. Tabulated data shown at Figures 13.2

Sample size n	Warning factor $A^1_{0.025}$	Action factor $A^1_{0.001}$
2	1·23	1·94
3	0·67	1·05
4	0·48	0·75
5	0·38	0·59
6	0·32	0·50
7	0·27	0·43
8	0·24	0·38
9	0·22	0·35
10	0·20	0·32

Fig 13.2 Factors used in $\bar{x}$ charts

and 13.3 is used (ref. BS 2564), this making the calculation of the standard deviation of the distribution of means (s_n), and the calculation of the standard deviation of the distribution of ranges (s_w) unnecessary.

Using this data, the control limits can be calculated thus:

$\bar{x}$ Chart

$$\text{(Action)} \quad \text{U.C.L.} = \bar{X} + A^1_{0.001}\,\bar{w}$$
$$\text{(Action)} \quad \text{L.C.L.} = \bar{X} - A^1_{0.001}\,\bar{w}$$
$$\text{(Warning)} \quad \text{U.C.L.} = \bar{X} + A^1_{0.025}\,\bar{w}$$
$$\text{(Warning)} \quad \text{L.C.L.} = \bar{X} - A^1_{0.025}\,\bar{w}$$

Solution

From (5)

$$p_B = m\pi \cos \psi$$
$$= 4\pi \cos 20°$$
$$= 4\pi \times 0.9397$$
$$\therefore \qquad p_B = 11.83 \text{ mm}$$

(iv) Measurement of base tangent. This is a measure of tooth spacing. Any calliper-type instrument or gauge can be used to *check base tangent*. A vernier calliper is often used, as shown at Figure 11.8, but commercial instruments, known as tangent comparators, offer the most convenient and accurate method.

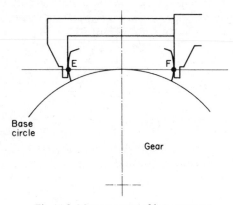

Fig 11.8 Measurement of base tangent

Comparison between Figures 11.4 and 11.8 shows base tangent *EF* being measured.

Example 11.4

Calculate the base tangent for a 20° pressure angle gear, having 45 teeth and a module of 4.

Solution

From (6)

$$t_B = p_B[0.0047T + K - 0.5]$$

where

$$p_B = 4\pi \cos 20°$$
$$= 11.83 \text{ (See Example 11.3.)}$$

and

$$K = 5 \text{ from table [See Section 11.1(iv).]}$$

'w' Chart

$$(Action) \quad U.C.L. = D^1{}_{0.999} \; \bar{w}$$
$$(Action) \quad L.C.L. = D^1{}_{0.001} \; \bar{w}$$
$$(Warning) \; U.C.L. = D^1{}_{0.975} \; \bar{w}$$
$$(Warning) \; L.C.L. = D^1{}_{0.025} \; \bar{w}$$

Let us take one example in order to illustrate the procedure.

Sample size n	Upper action factor $D^1{}_{0.999}$	Upper warning factor $D^1{}_{0.975}$	Lower warning factor $D^1{}_{0.025}$	Lower action factor $D^1{}_{0.001}$
2	4·12	2·81	0·04	0·00
3	2·99	2·17	0·18	0·04
4	2·58	1·93	0·29	0·10
5	2·36	1·81	0·37	0·16
6	2·22	1·72	0·42	0·21
7	2·11	1·66	0·46	0·26
8	2·04	1·62	0·50	0·29
9	1·99	1·58	0·52	0·32
10	1·93	1·56	0·54	0·35

Fig 13.3 Factors used in 'w' charts

Example 13.1

Samples of 5 were taken at regular intervals from a process, 10 samples in all being taken. The results were as follows:

Sample No.	Measurements per sample (hundredths of one mm)				
1	747	748	747	749	748
2	748	749	750	748	749
3	749	748	750	748	749
4	749	749	750	750	751
5	749	749	750	750	751
6	749	750	751	749	750
7	750	750	751	751	750
8	751	750	750	750	752
9	751	751	752	751	751
10	751	752	752	753	751

Calculate warning and action limits for $\bar{x}$ and 'w' Charts. Show the results of these samples plotted upon the charts.

Solution

Using an electronic calculator for speed the grand mean $\bar{X}$ of all 10 samples is found to be 7·499 mm, and the mean range $\bar{w}$ is found to be 0·018 mm.

$\bar{x}$ Chart

>   (Action) U.C.L. = 7·499 + (0·59 × 0·018) = 7·510 mm
>   (Action) L.C.L. = 7·499 − (0·59 × 0·018) = 7·488 mm
>   (Warning) U.C.L. = 7·499 + (0·38 × 0·018) = 7·506 mm
>   (Warning) L.C.L. = 7·499 − (0·38 × 0·018) = 7·492 mm

'w' Chart

>   (Action) U.C.L. = 2·36 × 0·018 = 0·042 mm
>   (Action) L.C.L. = 0·16 × 0·018 = 0·003 mm
>   (Warning) U.C.L. = 1·81 × 0·018 = 0·033 mm
>   (Warning) L.C.L. = 0·37 × 0·018 = 0·007 mm

The charts and plotted results are shown at Figure 13.4.

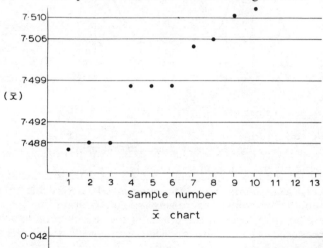

$\bar{x}$ chart

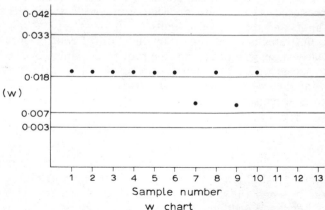

w chart

Fig 13.4 $\bar{x}$ and 'w' charts showing plotted results

13.4 Relative precision index

Nothing has yet been said about the relationship between the statistical control limits, and the tolerance called for in the specification. This is shown at Figure 13.5 where two examples are given:

(a) A component having a wide dimensional tolerance being manufactured by a process having little variability.
(b) A component having a narrow dimensional tolerance being manufactured by a process having great variability.

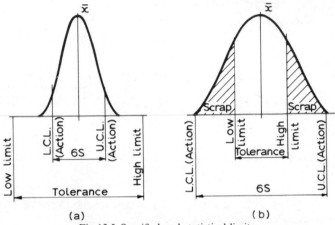

(a)   (b)

Fig 13.5 Specified and statistical limits

In case (a), the process is in effect too good for the specification. The statistical action limits have little meaning. It is possible to greatly vary $\bar{x}$ and 'w' without transgressing the specification limits.

In case (b), the process is too crude for the specification, much scrap being produced. Again, there is a wide divergence between the specification limits and the statistical limits.

This illustrates another important function of control charts, viz., they clearly show whether or not the process capability is compatible with the specification. The relationship between the two can be evaluated using the *relative precision index* (R.P.I.) which can be defined as the ratio of specification tolerance to average sample range, i.e.,

$$\text{R.P.I.} = \frac{\text{total specification tolerance}}{\text{average range}} = \frac{t}{w}$$

The values of R.P.I. are described as low, medium or high, depending upon the size of the sample, 'n'. Thus when 'n' = 4, for example, the R.P.I. is low if <3; medium if >3 or <4; high if >4.

The values of R.P.I. can be interpreted thus:

Low R.P.I.: with the range in control, rejections are inevitable even when the sample average is in control;

Medium R.P.I.: with the range in control, manufacture will be satisfactory providing the sample average is in control;

High R.P.I.: parts are being produced with little variation. This is the class of manufacture often associated with high precision machine tools.

The classification of R.P.I. for sample sizes up to 6 is given in the table shown at Figure 13.6.

Sample size	Low R.P.I.	Medium R.P.I.	High R.P.I.
(*n*)			
2	Less than 6	6 to 7	Greater than 7
3	Less than 4	4 to 5	Greater than 5
4	Less than 3	3 to 4	Greater than 4
5 and 6	Less than 2·5	2·5 to 3·5	Greater than 3·5

Fig 13.6 Relative precision index values

Example 13.2

The part produced in Example 13.1 has an average range of 0·018 mm and a tolerance of 0·03 mm. Calculate the R.P.I. (*n* = 5).

Solution

$$\text{R.P.I.} = \frac{\text{total specification tolerance}}{\text{average range}}$$

$$= \frac{0·03}{0·018} = 1·7$$

Therefore the R.P.I. value is low, being less than 2·5. This means that the process is unsatisfactory, and many rejects will be found in the samples.

13.5 Control charts for attributes

This type of *control chart* is used for manufactured articles which can be classed as good or bad, e.g., machined metal parts which are being inspected with limit gauges and being classified as acceptable or defective. The work is random sampled regularly, the results being plotted upon the control chart. The main purpose of the chart is to indicate changes in quality so that adjustments can quickly be made to correct the process before large quantities of scrap are produced, i.e., in effect the chart gives advance warning of the commencement of a trend towards the production of an increasing number of defective articles. There are two types of chart commonly used for the *control of attributes*, these being Fraction defective (*P*) charts, and Number defective (*m*) charts respectively. Each will be considered in turn.

Fraction defective charts. The compilation of these charts is perhaps best considered in two steps:

Step 1
Collect data from the process by taking samples of size '*n*' and counting the number of defectives in each sample, then calculate the fraction defective which = '*P*' = $\dfrac{\text{number of defects in sample}}{n}$

Draw up a frequency distribution of '*P*' values which should be approximately normal having a mean value of:

$$\bar{P} = \frac{\text{total number of defects found}}{\text{total number of parts sampled } (\Sigma n)}$$

Step 2
Turn the '*P*' distribution through 90°, and draw the inner control limits (warning), and the outer control limits (action) on the control chart. These limits are positioned at two standard deviations (2*s*), and three standard deviations (3*s*) respectively from the mean ($\bar{P}$). The standard deviation for a '*P*' distribution is calculated thus:

$$s = \sqrt{\frac{\bar{P}\,(1 - \bar{P})}{n}}$$

The resulting chart is called a '*P*' chart.

Consider an example which illustrates the construction of a '*P*' chart.

Example 13.3

Twenty samples of parts were taken from a production line for gauging, each sample containing 100 parts. The following number of defects were found in each sample:

3, 4, 5, 5, 5, 5, 4, 7, 6, 7,
6, 6, 4, 3, 5, 8, 5, 4, 6, 5.

(a) calculate $\bar{P}$ and draw the distribution, (b) calculate the limits and draw the control chart.

Solution

(a) $$\bar{P} = \frac{\text{total number of defects}}{\text{total number sampled}} = \frac{103}{2000} = 0.052$$

The distribution is shown at Figure 13.7.

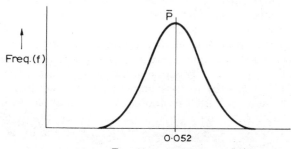

Fig 13.7 Distribution of '*P*' values

(b) $$s = \sqrt{\frac{\bar{P}(1 - \bar{P})}{n}} = \sqrt{\frac{0.052 \times 0.948}{100}}$$

$$= \sqrt{0.0005} = 0.022$$

∴ $2s = 0.044$ and $3s = 0.066$

(Action) U.C.L. $= 0.052 + 0.066 = 0.118$
(Action) L.C.L. $= 0.052 - 0.066 = 0$ (effectively)
(Warning) U.C.L. $= 0.052 + 0.044 = 0.096$
(Warning) L.C.L. $= 0.052 - 0.044 = 0.008$

The control chart is shown at Figure 13.8. In practice, the

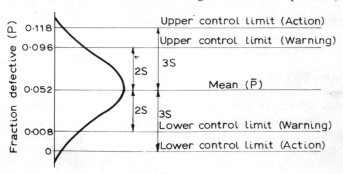

Fig 13.8 '*P*' chart

distribution curve is not drawn upon the chart, only the limits being drawn, as the chart at Figure 13.9.

Regular, random samples are taken by the inspector, and checked, the 'P' value for each sample being calculated and plotted upon the chart. We would expect (from our knowledge of the normal probability distribution) 99·73% of the points plotted to fall within the outer limits, and 95·45% to fall within the inner limits, due to the natural variability of the process. To put it another way, 2 points in 1000 should fall outside the outer limits, and 45 points in 1000 should fall outside the inner limits, by the laws of probability. If this is so, the process is deemed to be in statistical control. If more points start to fall outside the limits than can be attributed to chance, then the warning should be taken followed by action (if necessary) in order to correct the process. This means that an assignable cause for the drift outside the limits must be found, such as tool wear, for example. Points falling below the lower limits indicate an improvement in quality because the trend is then towards less rejects being found, so the lower limits are not as important as the upper.

Example 13.4

After the control chart in Figure 13.8 is compiled, the following defects from the process for 10 samples of 100 are obtained. Plot the results on the chart.

                3, 4, 4, 3, 4, 5, 6, 10, 11, 11, defects respectively.

Solution
Convert the results of sampling into 'P' values, where

$$`P' = \frac{3}{100}, \frac{4}{100}, \frac{4}{100}, \text{etc.}$$

The results are then:

        0·03, 0·04, 0·04, 0·03, 0·04, 0·05, 0·06, 0·10, 0·11 and 0·11.

These are shown plotted on the chart at Figure 13.9.

The process shows a drift towards running out of statistical control which if continued would require action, thus saving an accumulation of scrap. If the original sample results are plotted, then these will be seen to be in statistical control with the normal amount of variation inherent in the process Control charts exclude guesswork, leading to positive action being taken only when required.

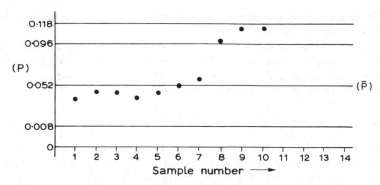

Fig 13.9 '*P*' chart showing plotted results

Number defective charts. The procedure is similar to that described for '*P*' charts, except that the limits can be read directly from a table.

Step 1
Collect data from the process by taking samples at random, and counting the number of defectives in each sample, then calculate the mean number of defects per sample ($\bar{m}$), where:

$$\bar{m} = \frac{\text{total number of defects found}}{\text{total number of samples}}$$

Step 2
Determine the upper control limits from the table at Figure 13.10. The resulting chart is called an '*m*' chart. As stated earlier, lower limits are of no great importance in the context of control charts for attributes. It will be seen that the table at Figure 13.10 only covers $\bar{m}$ values up to a maximum to 2·0. For values higher than this, one needs to use a Poisson distribution cumulative probability chart.

Average number of defectives expected in the sample ($\bar{m}$)	U.C.L. (Warning)	U.C.L. (Action)
0·6	2·5	3·8
0·8	2·9	4·4
1·0	3·3	4·8
1·2	3·7	5·2
1·4	4·0	5·6
1·6	4·4	6·1
1·8	4·7	6·5
2·0	5·0	6·8

Fig 13.10 Limits for number defective control charts

The following simple example shows how an '*m*' chart is constructed.

Example 13.5

(*a*) The results of 19 random samples for number of defective items is as follows:

0, 2, 2, 3, 2, 1, 1, 3, 2, 2, 4, 3, 3, 2, 4, 1, 0, 1, 2.

Determine the limits.

(*b*) The results of a further 25 samples of number defective taken subsequently are as follows:

2, 2, 2, 0, 1, 3, 1, 4, 2, 2, 3, 3, 0, 1, 2, 2, 3, 2, 0, 2, 6, 1, 5, 1, 4.

Plot these results upon the '*m*' chart.

Solution

(*a*)

$$\bar{m} = \frac{\text{total number of defects}}{\text{total number of samples}}$$

$$= \frac{38}{19} = 2 \cdot 0$$

From table at Figure 13.10

(Action) U.C.L. = 6·8
(Warning) U.C.L. = 5·0

(*b*) The control chart is shown at Figure 13.11 with the plotted results.

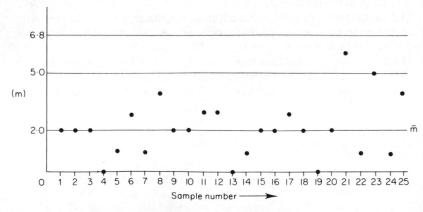

Fig 13.11 '*m*' Chart showing plotted results

13.6 Interpretation of results

Control charts, of course, show the pattern of variability of a process, and it should again be remembered that one expects 2 points in 1000 to fall outside the outer (action) limits, and 45 points in 1000 to fall outside the inner (warning) limits because of chance causes. Bearing this in mind, the *information* shown on the control chart has to be *interpreted* with a view to deciding whether or not the results reveal an unstable production situation, i.e., is the process out of control. BS 2564 gives useful guidelines, viz., manufacture should be considered unstable, or out of control, when:

(i) one point appears in close proximity to, or outside the outer limits; or

(ii) two points in any consecutive ten appear in close proximity to, and/or outside the inner limits; or

(iii) three points in any consecutive twenty appear in close proximity to, and/or outside the inner limits.

It should be emphasised here that a lot of experience is needed in correctly interpreting the information recorded on control charts, and before drastic and precipitous action is taken it is always wise to take more samples in order to consolidate the results.

Learning Objectives 13
The learning objectives of this section of the TEC unit are:

13. Understands and constructs control charts for variables and attributes.

13.1 Lists the criteria which justify the application of statistical control of quality techniques.

13.2 Distinguishes between assignable and random causes of variation.

13.3 Constructs control charts for sample average and sample range to monitor process variation.

13.4 Estimates a suitable relative precision index from derived data.

13.5 Constructs fraction defective and number defective control charts.

13.6 Interprets information derived from charts constructed in 13.3 and 13.5.

Exercises 13

1 List the criteria which justify the application of statistical control quality techniques.

2 What is meant by (i) assignable causes of variation, and (ii) random causes of variation? Give examples.

3   Twenty samples of four components per sample were taken from a
process, and the length of the parts measured in metres. The results
are as follows:

Sample No.

1	1·16	1·25	0·66	0·56
2	0·84	0·82	0·92	0·60
3	0·97	0·94	0·99	0·90
4	1·00	0·94	1·50	1·18
5	0·75	0·97	0·47	0·73
6	0·92	0·60	0·82	1·14
7	1·17	1·00	0·85	0·36
8	0·68	0·93	0·89	1·13
9	1·00	0·91	0·60	0·68
10	0·97	0·87	0·71	0·89
11	0·73	0·66	0·79	0·59
12	0·82	0·77	0·67	0·70
13	0·90	1·25	1·00	0·81
14	0·57	0·62	0·61	0·69
15	0·61	1·02	1·45	0·93
16	0·81	1·00	1·25	0·90
17	0·71	0·94	0·87	0·84
18	0·97	1·06	1·10	0·89
19	1·12	0·73	0·62	0·78
20	0·68	0·61	1·00	1·11

Draw $\bar{x}$ and 'w' charts. Plot the given results upon the charts, and
determine whether or not the process is in statistical control.
(Ans. $\bar{x}$ Chart

$$\text{U.A.L.} = 1\cdot17 \quad \text{L.A.L.} = 0\cdot56$$
$$\bar{X} = 0\cdot864 \ \text{U.W.L.} = 1\cdot05 \quad \text{L.W.L.} = 0\cdot67$$

'w' Chart

$$\bar{w} = 0\cdot406 \quad \text{U.A.L.} = 1\cdot05 \quad \text{L.A.L.} = 0\cdot04$$
$$\text{U.W.L.} = 0\cdot79 \quad \text{L.W.L.} = 0\cdot12.)$$

4   If $\bar{w}$ for samples of three is 25 mm, and the specification tolerance is
140 mm, calculate the R.P.I. and indicate its category. (Ans. R.P.I.
= 5·6.)

5   Samples of 200 components taken at random each day, are gauged
as a means of checking. The following results are given for the first
24 days, in terms of fraction defective, ('P').

0·45, 0·5, 0·45, 0·6, 0·35, 0·55, 0·65, 0·75, 0·75, 0·65, 0·45, 0·55,
0·5, 0·6, 0·35, 0·5, 0·45, 0·45, 0·5, 0·6, 0·55, 0·55, 0·6 and 0·5.

(a) From this sample, calculate $\bar{P}$, and the position of the action
and warning limits.

(b) Draw a 'P' chart, and plot the results of the 24 days sampling upon the chart. Is the process in statistical control for this period of production?

(Ans. (a) $\bar{P}$ = 0·53.   U.W.L. = 0·60,   L.W.L. = 0·46
U.A.L. = 0·635,   L.A.L. = 0·425)

6   25 samples of 20 components were taken at random from a process, and the number of defectives in each sample were:

2, 2, 1, 0, 2, 1, 4, 4, 5, 2, 1, 2, 1, 0, 1, 3, 3, 0, 2, 1, 2, 0, 0, 5, 0.

Derive the limits, and draw up a control chart for number defective.

Plot the above results upon the chart.

(Ans. $\bar{m}$ = 1·8   U.A.L. = 6·5
U.W.L. = 4·7.)

7   Explain a sensible policy for interpreting the information recorded on control charts, and acting upon it.

Chapter 14
Jigs and Fixtures

The practice of designing and making jigs and fixtures has been in wide use for many years, and the number of these devices successfully used is legion. We are therefore considering a vast subject in which a great deal of personal preference, variety, inventiveness and even engineering dogma may be found. Therefore, the student of jig and fixture design may be puzzled at first sight by the apparent lack of uniformity. However, certain sound design principles are now well established and will shortly be discussed, and also much standardisation has been applied in this field of production engineering leading to cheaper production methods.

14.1 Justifying the need for jigs and fixtures

In engineering there is often more than one way of manufacturing a component, and *cost* is often the *criterion* of choice of method. If a component is required in large quantities, then clearly a method which is suitable for producing one-off (such as marking out, setting on machine, clamping to machine table, etc.) would not be suitable for economic reasons. A faster and more profitable method requires some device on which the component(s) can quickly be position in the correct relationship to the cutting tool(s) and quickly clamped before machining takes place. Such a device is known as a jig or fixture. Hence, jigs and fixtures ensure that the component has metal removed from it, by machining in the correct place. They make it possible to machine at much greater speed, and with greater accuracy than is possible using hand methods. Responsibility for the accuracy of the finished part is taken from the operator, and is instead a function of the jig or fixture. The number of parts being produced in the production run has to be high enough to justify the cost of the jig or fixture. Figure 14.1 shows that once the production quantity has passed the break-even quantity Q_E, then subsequent production costs, using the jig or fixture, are less. It should be remembered that the labour costs of production are less using jigs and fixtures, as skilled

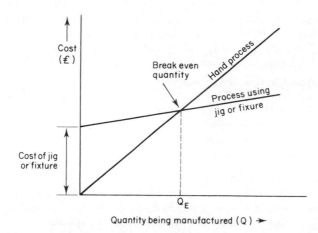

Fig 14.1 Break-even cost chart

workers are not required. Section 6.3 gives a brief description of a break-even cost analysis, and as explained there, cost is usually the major criteria to be considered when alternative methods of manufacture are being examined.

The concept of interchangeability can only be maintained by the consistent production of components to the required degree of accuracy. The use of jigs and fixtures gives this essential consistency at an acceptable speed and cost.

14.2 Difference between jigs and fixtures

A *jig* is a device usually made of metal which locates and holds the workpiece(s) in a positive manner and also guides the cutting tool(s) such that it is in the correct relationship to the work when machining commences. It is usually necessary for the work to be held in the jig by clamping. The jig is not fixed to the machine table by clamping but is held by hand. Jigs are used for quantity drilling, reaming and tapping for example.

A *fixture* is a device similar to a jig but as the name implies is fixed to the machine bed by clamping in such a position that the work is in the correct relationship to the cutter. A further difference is that the cutter is not guided into position ready for machining to commence. A setting gauge is often provided to enable the initial setting of work to the cutter to be quickly and easily accomplished before production begins. Fixtures are used for quantity milling, turning and grinding for example.

These definitions are not always precisely applied in engineering, and the terms jig or fixture are often used quite loosely. The terms also cover all classes of devices from a simple clamp and location stop attached to a machine table, to a fully automatic, power operated, multi-headed fixture of the most complex and sophisticated design. Again, cost and quantity will usually be the deciding factors in the evolution of the jig or fixture.

14.3 Jig and fixture design features

The basic principles of location and clamping, which are fundamental to the *design* of all *jigs* and *fixtures*, will be considered in this section. This will be supplemented by showing the application of these principles to particular processes, viz., drilling, milling, turning, grinding, boring and welding. A broaching fixture is shown at Figure 6.6.

Principles of location. As discussed in Section 9.1 kinematics is the branch of mechanics relating to problems of motion and position, and kinematic principles can be applied in considering the location of work in a jig or fixture. Figure 9.1 shows a body in space.

It can be seen that the body has six degrees of freedom. It can rotate about, or have linear movement along each of three axes, XX, YY or ZZ. It is not constrained or prevented from movng in any direction. When located in a jig a workpiece must be constrained from moving in any direction. This can be done by six locations in the case of the body shown in Figure 9.1, this being known as the six point location principle. It is illustrated in Figure 14.2.

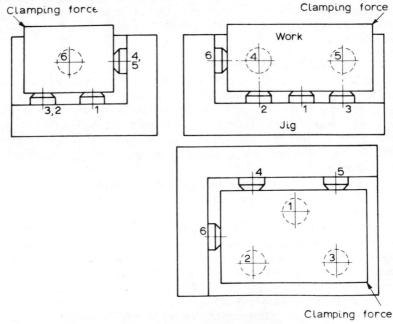

Fig 14.2 Six-point location principle

The base of the component is resting on three location pegs which is the minimum number of points upon which it will firmly seat. When the closing force is applied (by means of a clamp say) all the six degrees of freedom have been removed. If more than six points are used the additional points will be surplus and unnecessary and would therefore be redundant constraints.

Depending upon the shape of the component six location pegs may not be required in practice and may be replaced by other devices. However the same basic principle prevails. Three examples of alternative means of location are given next, each of which has the six degrees of freedom removed without having redundant constraints.

(i) *Location upon two plugs.* The smaller hardened and ground plug has flats machined on each side to allow for slight variations in centre distance of the two holes in the lever. Two completely round location plugs would allow much less variation in the hole centres, thus providing a very precise form of location.

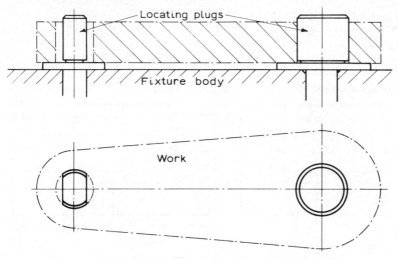

Fig 14.3 Location of lever upon two plugs

(ii) *Location in two vees.* This is a common form of location when the shape of the component will allow it to be used. With one vee sliding as shown it can also be used for clamping as well as locating. A combination of both location methods shown in Figures 14.3 and 14.4 can be used by locating the component upon a round plug at one end and in a sliding vee at the other end.

(iii) *Location upon a plug or in a spigot recess.* Depending upon the sequence of machining operations a component may have to be located

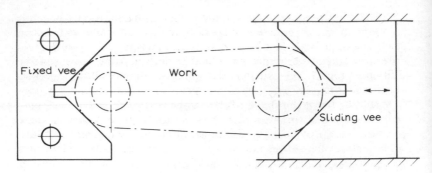

Fig 14.4 Location in two vees

internally upon a plug [Figure 14.5(*a*)], or located externally in a recess [Figure 14.5(*b*)].

Plugs should not be screwed into the fixture body as a threaded hole will not allow the plug to be positioned in the fixture with sufficient accuracy.

Spigot recesses may be cylindrical or of some particular profile to suit the shape of the component.

Design features to consider with respect to location

(*a*) Location influences the accuracy and quality of a component. Location should be designed to kinematic principles thus reducing the six degrees of freedom to zero with no redundant location features.

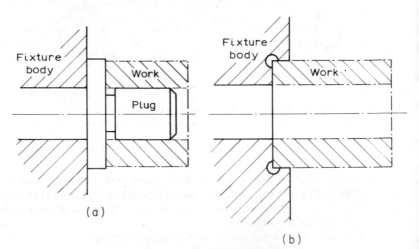

Fig 14.5 Plug and spigot recess location

(b) With first operation work upon rough unmachined surfaces use three-point location where possible, and adjustable or expanding locators to allow for large variations in size.

(c) Locate from the same machined surface (datum) for as many operations as possible in order to reduce the possibility of error.

(d) Make sure the location is 'foolproof', i.e., the component can only be loaded into the fixture in the correct position.

(e) Location features should not be swarf traps and should have clearance provided where necessary to clear the machining burrs.

(f) Consider the operator's safety and the manipulative difficulties in loading the component into the fixture. Retractable location pins may be necessary.

Principles of clamping. As with location, clamping will influence the accuracy and quality of the component and will further influence substantially the speed and efficiency of the operation being carried out on the component. Many types of clamps and clamping methods have been standardised and are suitable for the great majority of engineering operations. The main principle to observe when designing any clamping arrangement is that the fixture, workpiece and location features should not be distorted or strained. This means that clamping

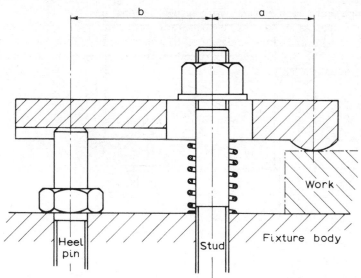

'a' should not be greater than 'b'

Mechanical advantage $MA = \dfrac{a+b}{b}$

Fig 14.6 Bridge clamp

forces should not be excessive but sufficient to hold the work rigidly and should be applied at points where the work has the support of the solid metal of the fixture body. Three examples of alternative means of clamping are given next, chosen from a great variety of standard examples.

(i) *Bridge clamp.* An example of a bridge clamp is given at Figure 14.6.

From Figure 14.6 it can be seen that as the nut is unscrewed the spring pushes the clamp upwards. The clamp has a longitudinal slot so that it can be pushed clear of the work. To speed up the clamping operation the hexagonal nut may be replaced by a threaded handle, or a quick-action locking cam.

(ii) *Two-way clamp.* This clamping system enables a two-way clamping action to be obtained from one nut or threaded handle. Clamping force is applied to the top and one side of the work-piece. The clamp has a quick-release action. As the nut is released, the hinged stud can be swung clear from the slot in the end of the top hinged clamp. More elaborate versions of this and other multi-way clamps have been in use for many years. Figure 14.7 shows such a clamp.

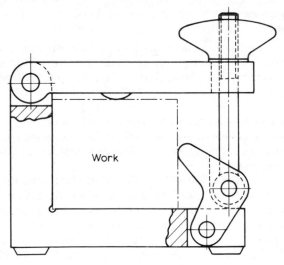

Fig 14.7 Two-way clamp

(iii) *Wedge operated clamp.* A wedge operated clamp is illustrated at Figure 14.8, reference to which shows that horizontal linear movement of the wedge clamps the workpiece vertically. Alternative manually operated devices for the wedge are a screw or cam. A further alternative is that the wedge could be operated by pneumatic cylinder or hydraulic cylinder. This then leads to the possibility of automatic

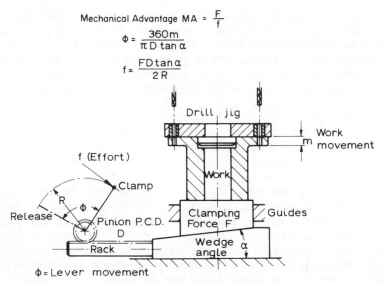

Mechanical Advantage MA $= \dfrac{F}{f}$

$$\Phi = \dfrac{360\,m}{\pi\,D\,\tan\alpha}$$

$$f = \dfrac{FD\tan\alpha}{2R}$$

Fig 14.8 Wedge operated clamp

clamping of the workpiece as part of a fully automatic machining cycle. Where air or fluid power is being used for actuating clamps, safety arrangements must be made to protect the operator from trapping his fingers or hand.

Design features to consider with respect to clamping

(a) Clamping influences the accuracy and quality of a component. Clamps should be applied to the component where it is rigid and well supported.

(b) Clamping forces should be controlled such that they are not great enough to distort the location, work, fixture or the clamps themselves. Note that the magnitude of cutting forces to be resisted varies with different machining operations.

(c) Clamps should be quick acting and as simple as possible. Some quick-acting devices are expensive and the costs should be considered to see that such devices are justified.

(d) Ensure that the clamps are well clear of cutting tools, and that adequate clearance is available with the clamps released to load and unload the work in safety. Ejectors may be necessary for ease of unloading.

(e) Sound ergonomic principles should be applied when designing the clamping system to be used by an operator. Note that ergonomics is the science of fitting the job to the worker.

Application of design principles. Many considerations, including technical and economic, will influence the jig and tool designer before the
final fixture design is evolved. The magnitude and importance of these
considerations will of course depend upon how simple or elaborate the
production process is to be. Despite standardisation nearly every fixture or jig is unique to some extent in design and creation. However,
whatever the process or quantities required, certain features are of
major importance in any fixture design. These are:
   (*a*) Method of locating the component(s) in the fixture.
   (*b*) Method of clamping the component(s) in the fixture.
   (*c*) Method of positioning the tool relative to the component.
   (*d*) Method of positioning the fixture relative to the machine. This
        is not required with a jig which is not clamped in some fixed
        position upon a machine table, but even so some form of stop may
        be provided against which the jig may be held to give a quick
        initial positioning.
   (*e*) Safety aspects to protect the operator whilst using the fixture.
   A complete text book is required to analyse all the great variety of
these major features as applied in various machining processes, but
one example of a jig or fixture for the common processes will be given
in order to illustrate good practice.

Drill jig

From Figure 14.9 the following design features should be noted.
   (*a*) Location is from a previously machined bore in the component
        on to a hardened and ground male spigot.
   (*b*) Clamping is against the spigot shoulder with a spider clamp
        plate, actuated by a cam operated plunger. The method adopted
        will not cause distortion. When the cam lever is pulled to the
        release position the spider clamp is easily and quickly removed
        facilitating rapid unloading of workpiece.
   (*c*) The cutting tool (twist drill) is positioned relative to the work by
        means of hardened and ground drill bushes which act as guides
        for the drill. Proportions for these bushes are given by British
        Standard Specification 1098. The gap between bush and work is
        large enough to avoid being a swarf trap, but not too large to
        give the drill inadequate support as it enters the work.
   (*d*) No method of positioning the jig relative to the drilling machine
        table is necessary as the jig is held by hand.
   (*e*) A guard should be provided around the drill. Adequate clearance
        is provided so that the swarf can easily fall clear through the jig
        body and spider clamp, and no swarf traps are present. Hence
        the danger of the drill breaking and causing injury is greatly
        reduced. The jig has no sharp corners.

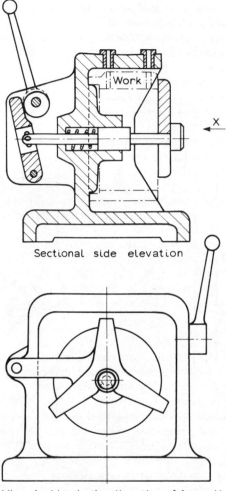

Sectional side elevation

View looking in the direction of Arrow X

Fig 14.9 Drill jig

Further features that form part of good drill jig design practice are:

(*f*) The jig which is handled frequently during use is of light construction consistent with rigidity. The jig body can be a casting as shown or alternatively it can be fabricated from screwed and dowelled steel plate.

(*g*) The axial cutting force, as is desirable, is not taken directly by the clamps. This should always be so although it is sometimes impossible to observe this general principle.

(*h*) The base of the jig is so designed that the feet cannot fall into the drilling machine table tee-slots.

(*i*) Foolproofing pins so arranged that the component cannot be placed in the jig other than in the correct position.

Milling fixture

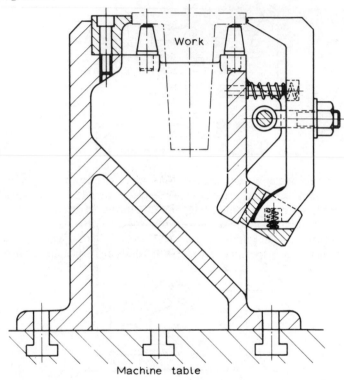

Fig 14.10 Milling fixture

From Figure 14.10 the following features should be noted.

(*a*) No precise location required as this is the first operation to be carried out on a casting. The component flange rests upon four hardened and ground pins (four being necessary in this case for stability) which have domed tops giving point contact on the uneven casting surface. With great variation of casting size, the pins would have to be made adjustable and capable of being locked in the required position.

(*b*) The clamping arrangement has to be below the top surface of the component which is to be face or slab milled. The side bridge clamp shown with its serrated edge pulls the component firmly down on to the pegs and against the opposite serrated jaw. The clamp is of the equalising type which allows for considerable variation in the casting size. This fixture, as with many milling fixtures, could be designed to hold more than one component in tandem by multiplying the number of clamping arrangements.

Milling cutting forces are high in magnitude, hence the clamps must be of robust design. Cam arrangements should be avoided as they can be loosened during cutting and hexagon nuts and spanners are preferable.

(c) Precise positioning of the milling cutter relative to the work is not necessary, but when required a setting block screwed and dowelled in the correct position on the fixture body can be used in conjunction with a feeler gauge. This is shown in Figure 14.11.

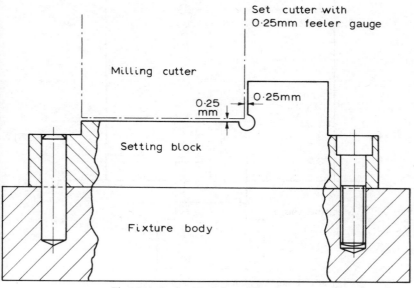

Fig 14.11 Setting block for milling fixture

(d) The fixture would require bolting to the machine table while machining takes place. Again precise positioning is not necessary in this case, but where it is necessary, tenon blocks screwed to the underside of the fixture can be used to locate the fixture on the machine table in the correct position. This ensures that the fixture is in line with the axis of the machine table. This is shown in Figure 14.12.

(*e*) Milling can be a very dangerous machining operation. The machine should never be started up without a guard around the cutter, and a limit switch could be arranged to ensure that this is always so. The release points for the clamps should always be well clear of the cutter.

Further features that form part of good milling fixture design practice are:

(*f*) The fixture is of heavy, rigid construction to withstand the high milling cutting forces. A cast iron casting is chosen for the fixture body for this reason and also because it has the property of damping the vibration present during the milling operation.

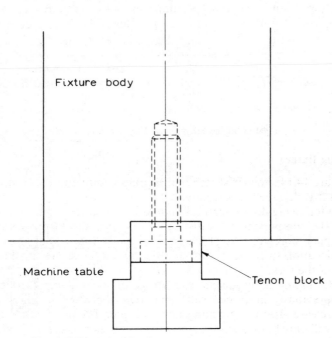

Fig 14.12 Tenon location between fixture and machine table

(*g*) With up-cut milling technique using a slab mill there is a component of the cutting forces tending to lift the component up out of the fixture. With the wedge-shaped serrated clamp and jaw used this is successfully counteracted.

(*h*) The work is set against a solid face to resist the considerable feed force of the milling operation. This prevents any possibility of the work being pushed out of the fixture. This is shown in Figure 14.13.

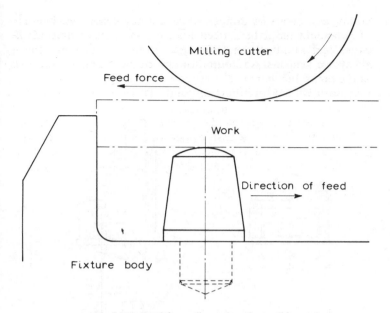

Fig 14.13 Feed force directed against solid metal

Turning fixture

Figure 14.14 shows a sketch of a turning fixture and the component which it is designed to accommodate.

The following design features should be noted.

(a) The component is located in a horseshoe slot (as opposed to a plain bore) for ease of loading and is also located flat against the location spigot to ensure that the bore is square to the flat faces of the fork.

(b) Clamping is by means of a bridge or latch clamp through the previously machined slot. The stud is pivoted to allow quick release after untightening nut. The clamp is so arranged that it will only fit through the slot if the component is correctly aligned.

(c) The location features of the fixture are so arranged that the drill and reamer will be correctly positioned relative to the work when they are fitted in the capstan lathe turret.

(d) The fixture is positioned relative to the machine by means of a location spigot on the back of the fixture body which fits the recess in the spindle flange on the machine.

(e) All parts of the fixture are within the outside diameter of the fixture body. Integral balance weights ensure that there are no out of balance forces which could cause danger (particularly at high speeds) to the operator. A guard must be provided which completely covers the fixture when it is revolving. The question of balancing the fixture when loaded is important and can be considered as a further feature apart from that of safety.

(*f*) A balance weight(s) is required to ensure the geometric accuracy of the hole when bored. The bored hole will not be truly round if machined with the fixture badly out of balance.

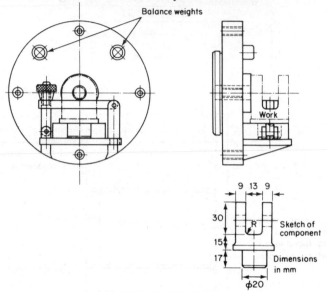

Fig 14.14 Turning fixture

Grinding fixture

A grinding fixture for a cylindrical grinding operation is shown at Figure 14.15.

There are no design features peculiar to grinding fixtures that have not already been covered. Cylindrical grinding fixtures are similar in principle to turning fixtures, and surface grinding fixtures are similar in principle to milling fixtures. Neither need be as robust since grinding cutting forces are comparatively light. Much simple work can easily be located and gripped on magnetic chucks with little risk of distortion.

In view of the above remarks, an expanding mandrel has been chosen as an example of a grinding fixture to add to the variety of the examples. The work is being form ground on a cylindrical grinding machine, and the gripping power of the expanding sleeve is adequate to resist the grinding forces. The design of the sleeve is such that it imparts a grip on all the surface area of the previously machined bore. It is particularly important with a finishing process such as grinding

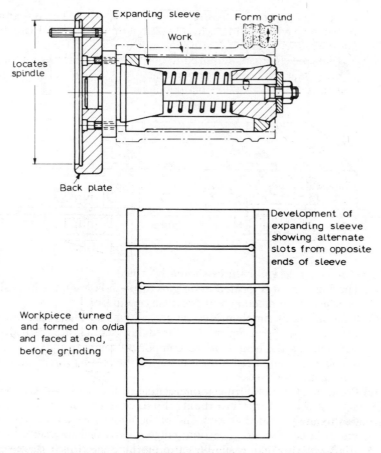

Fig 14.15 Grinding fixture

that the component does not move after being released from the clamps hence destroying the accuracy. For the same reason of maintaining accuracy, spindle spigot fits must be precise, balancing where required must be precise and rigidity is important to stop vibration during grinding.

Horizontal boring fixture

Figure 14.16 illustrates a horizontal boring fixture. Note that this type of holding device for boring, commonly called a fixture, is in effect a jig by definition, as the cutting tools may be guided from the jig. The horizontal boring process is ideal for components which would be difficult to machine on a lathe because of their size and shape.

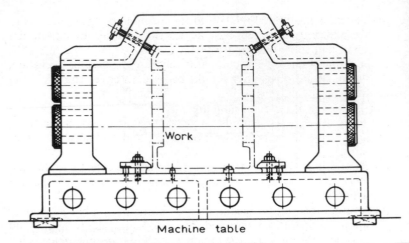

Fig 14.16 Horizontal boring fixture

The following design features should be noted.

(a) The component is located upon pins through holes in the base all of which were machined in previous operations.

(b) Clamping is by simple bridge clamps, access to which is provided through the open jig body. Steadying screws are provided overhead as the height of the component tends to make it unstable. Clamping need not be heavy as the cutting forces during boring are not excessive.

(c) Positioning of the boring tools relative to the work involves designing two features into the jig. First, close (running) fit steady bushes are provided at each end of the jig into which the boring bar is fitted. Second, a setting gauge is provided to ensure that the boring tool can easily be set to machine the correct diameter bore. See Figure 14.17.

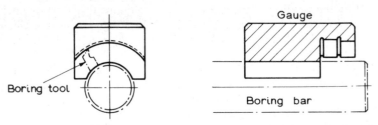

Fig 14.17 Setting gauge for boring tools

If a roughing and finishing tool is incorporated in the same bar then the gauge can be as shown.

(*d*) The jig is positioned relative to the machine by fitting tenons to the underside of the jig base as described earlier for milling fixtures, and shown at Figure 14.12.

(*e*) Horizontal boring is a comparatively safe machining operation, but quick-fitting guards should be used to cover the revolving tools during cutting.

Further design features to note are:

(*f*) Adequate clearance should be provided at each end between the jig and the work to allow easy setting and locking of the tool bits.

(*g*) The boring bar should be easy to fit and release from the machine spindle nose as this has to be done every time a component is bored.

(*h*) Several tools can be accommodated in the boring bar depending upon the operations required. These tools include roughing and finishing boring tools, shell reamers, facing tools and chamfering tools.

Welding fixture

A fixture in which parts are assembled ready for electric arc welding is shown in Figure 14.18.

Welding fixtures can be classed as assembly and welding fixtures. A typical operation is that of locating and clamping several components in a fixture ready for welding into a fabricated assembly. The process may be gas or electric arc welding, spot or seam resistance welding, or even brazing or soldering.

The location and clamping principles described earlier for machining operations should be adhered to, but there are some design features applicable only to welding fixtures which should be considered. These are:

(*a*) Expansion will take place due to the great local heat generated by the welding process. Should the component distort or buckle as a result, the locations should be so designed that the welded assembly does not lock in the fixture hence preventing its removal.

(*b*) The fixture should be rigid and stable for the same reason. In aircraft assembly work, fixtures of this type may be of enormous size, and tubular fabricated construction is often preferred for these fixtures giving great strength with lightness.

(*c*) Clamps should be arranged in such a way that they do not distort the component. Heavy clamping pressure is often unnecessary and finger screw operated clamps or toggle clamps are frequently used. On car body assembly lines many examples of toggle clamping can be seen, much of it pneumatically operated.

(d) Weld spatter can be a problem where electric arc welding is used particularly if it adheres to clamp screws. Where possible clamps should be kept clear of the welding zone, or should be shielded against the process.

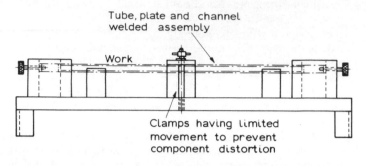

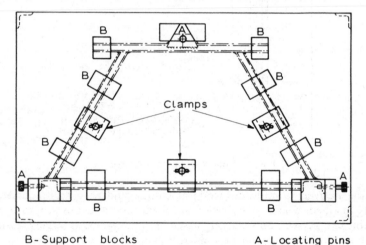

B- Support blocks   A- Locating pins

Fig 14.18 Welding fixture

(e) The location and clamping arrangement should not be so elaborate, that the welding zone is inaccessible to the welding operator. Accessibility and simplicity are important factors.

(f) In small, spot-welded assemblies, a simple nutcracker type of jig is often adequate. This can be held in the hand, the work being presented under the electrodes by the operator, and the position of the weld 'slug' being judged by eye. If the spacing of the slugs is important then an indexing fixture can be used with a single pair of electrodes. See Figure 14.19.

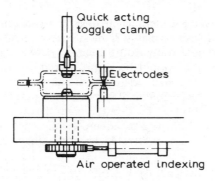

Quick acting
toggle clamp

Electrodes

Air operated indexing

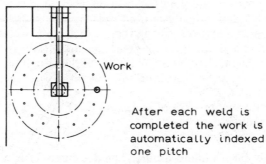

Work

After each weld is
completed the work is
automatically indexed
one pitch

Fig 14.19 Spotwelding indexing fixture

Alternatively, a multi-electrode machine could be used if economically justified, where many welds are made simultaneously. Examples of this can be seen on car body assembly lines where a large pressing such as a floor panel can be welded very quickly to the body.

14.4 Design analysis

Before any fixture design is finalised all of the factors mentioned previously, both economic and technical, have to be considered. This is done in a *design study* where each of the following steps are taken in turn:

   (i) Study the component drawing paying particular attention to the primary process (such as forging, die casting, etc.) and the tolerances on dimensions laid down by the designer.
   (ii) Study the planning sheet in order to ascertain the manufacturing sequence of operations laid down by the planning engineer.
   (iii) Study in detail the operation for which the fixture and tools are

required, and decide by which process or processes the operation can be carried out. It will be necessary to estimate the standard time, where standard time can be defined as: 'the total time in which a job should be completed at standard performance'. In turn, standard performance can be defined as: 'the rate of output which qualified workers will naturally achieve without over-exertion as an average over the working day or shift provided they know and adhere to the specified method and provided they are motivated to apply themselves to the work'.

(iv) Carry out a costing exercise in order to discover which of the technically feasible alternative processes is economically viable. Use the break-even analysis described in Section 14.1 for this purpose.

(v) Complete the design study of the fixture by outlining the methods to be adopted for location, clamping, tool positioning, etc. This design analysis need only include sketched details of the design features showing their principle of operation.

The work done in this part of the design study is passed over to the tool design draughtsman who will complete the design on the drawing board and produce finished working drawings with each individual item detailed.

To illustrate this procedure, an example is now given of a design study for a milling fixture.

Design study for a milling fixture. The design study is required for a fixture for the machining of the faces marked thus: $\triangledown$ on the component shown in Figure 14.20.

The component shown is required in weekly batches of at least 3000, the total lot size being substantial. The following points should be considered:

(i) The component is a casting with a mould parting line and some draft or taper which might warrant attention during location and clamping.

(ii) The sequence of operations is such that the bore and each end face of the 32 mm diameter boss have already been machined. This is to be followed by the machining operation on the pads.

(iii) Assuming a 40-hour week, the minimum production rate is
$$\frac{3000}{40} = 75/\text{hour}.$$

The pad faces could be machined by horizontal or vertical milling, or surface broaching. Assume that a horizontal milling machine and a surface broaching machine are available. It will now be necessary to check if these machines can achieve a production rate of 75/hour.

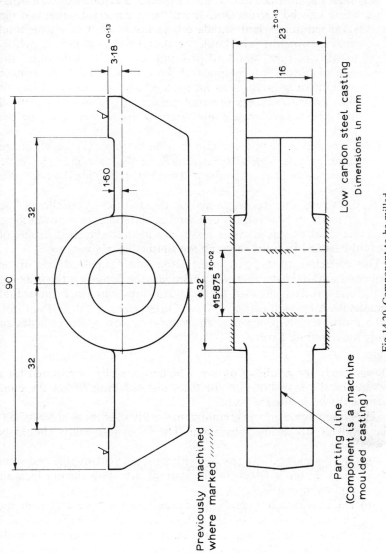

Previously machined
where marked //////

Parting line
(Component is a machine
moulded casting)

Low carbon steel casting
Dimensions in mm

Fig 14.20 Component to be milled

Milling (See Figure 14.21). Assume a 100 mm diameter cutter having 12 teeth is used and the depth of the cut is 3 mm. A suitable cutting speed is 305 mm/s at a feed of 0·08 mm/tooth.

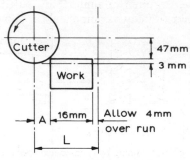

Fig 14.21 Milling operation

From Figure 14.21, it can be shown that:

$$A = \sqrt{d(D - d)}$$

where A = Approach distance; D = Cutter diameter; d = Depth of cut.

Hence, $A = \sqrt{d(D - d)} = \sqrt{3 \times 97} = \sqrt{291} = 14·15$ mm.

Also $L = 16 + A +$ over-run $= 16 + 15 + 4 = 35$ mm.

Cutter speed $= \dfrac{305}{100\pi} = 0·97$ rev/s: 0·90 rev/s is available on the machine.

At a feed of 0·08 mm/tooth, feed/rev $= 0·08 \times 12 = 0·96$ mm.

If f = feed in mm/s
then $f = 0·96 \times 0·90 = 0·864$ mm/s: 0·80 mm/s is available on the machine.

$\therefore$ Time $T = \dfrac{L}{f} = \dfrac{35}{0·80} = 44$ seconds.

By synthesis the unloadings, loading time, etc., is found and added to a relaxation allowance giving a total allowance of 34 seconds.

$\therefore$ Standard time $= 44$ s $+ 34$ s $= 78$ s $= 1·30$ min.

$$\text{Output} = \frac{60}{1\cdot30} = 46 \text{ components/hour.}$$

This is below the required minimum rate but it is found that the machine will accommodate two components as shown in Figure 14.22.

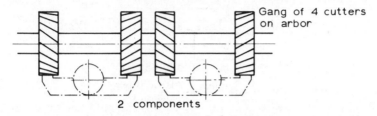

Gang of 4 cutters on arbor

2 components

Fig 14.22 Milling two components per cycle

Therefore the fixture could be designed to hold two components.

The allowances to the machining time are now found to be increased from 34 s to 49 s

$\therefore$ Standard time $= 44\,\text{s} + 49\,\text{s} = 93\,\text{s} = 1\cdot55$ min.

$$\text{Output} = \frac{60}{1\cdot55} \times 2 = 77 \text{ components/hour.}$$

This is satisfactory.

Broaching (See Figure 14.23). Assume a cut/tooth of 0·025 mm and a cutting speed of 50 mm/s.

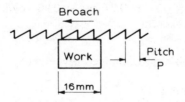

Broach

Work

Pitch
P

16 mm

Fig 14.23 Broaching operation

From Figure 14.23:

Pitch $P = 1\cdot77 \sqrt{\text{length of cut}} = 1\cdot77 \sqrt{16} = 1\cdot77 \times 4$
$= 7\cdot08$ say 7 mm.

Number of teeth required $= \dfrac{3}{0\cdot025} = 120$ teeth

$\therefore$ Length of cutting portion required $= 120 \times 7 = 840$ mm.

$\therefore$ Time $T = \dfrac{840 + 16}{50} = \dfrac{865}{50} = 17 \cdot 1$ seconds.

Using the same allowance as before

$$\text{Standard time} = 17 \cdot 1\,\text{s} + 34\,\text{s} = 51 \cdot 1\,\text{s} = 0 \cdot 85 \text{ min.}$$

$$\text{Output} = \frac{60}{0 \cdot 85} = 71 \text{ components/hour.}$$

With a little increase in cutting speed this can easily be raised to 75 components/hour. This is satisfactory.

   (iv) The milling fixture will be more expensive than the broaching fixture as it must be designed to hold two components. On the other hand the broach will be much more expensive than the milling cutters which could possibly be a stock item. An estimate must be made here of the various tooling costs involved (including the cost of the fixture) in order to make a reasonably accurate break-even analysis. Past costing records which are accurate can be of enormous help here.

	Milling Process	*Broaching Process*
(*a*) Tooling cost (estimated)	£450	£500
(*b*) Material cost/component	£0·72	£0·72
(*c*) Operating labour cost	£1·44/h	£1·44/h
(*d*) Standard time/component	0·78 min.	0·85 min.
(*e*) Setting up labour cost	£2·88/h	£2·88/h
(*f*) Setting up time	2h	4h
(*g*) Machine overheads	300% of (*c*)	500% of (*c*)

Milling Machine

$$\text{Overheads} = \frac{300}{100} \times 1 \cdot 44 = £4 \cdot 32/\text{h}$$

$$\text{Fixed cost} = 450 + 2(2 \cdot 88 + 4 \cdot 32)$$
$$= 450 + 14 \cdot 40 = £464 \cdot 40$$

$$\text{Variable cost/component} = \left(1 \cdot 44 \times \frac{0 \cdot 78}{60}\right) + 0 \cdot 72 + \left(4 \cdot 32 \times \frac{0 \cdot 78}{60}\right)$$

$$= \left(5 \cdot 76 \times \frac{0 \cdot 78}{60}\right) + 0 \cdot 72 = £0 \cdot 7949$$

Variable cost/1000 components $= £794 \cdot 90$

Broaching Machine

$$\text{Overheads} = \frac{500}{100} \times 1\cdot44 = £7\cdot20/\text{h}$$

$$\text{Fixed cost} = 500 + 4(2\cdot88 + 7\cdot20)$$
$$= 500 + 40\cdot32 = £540\cdot32$$

$$\text{Variable cost/component} = \left(1\cdot44 \times \frac{0\cdot85}{60}\right) + 0\cdot72 + \left(7\cdot20 \times \frac{0\cdot85}{60}\right)$$

$$= \left(8\cdot64 \times \frac{0\cdot85}{60}\right) + 0\cdot72 = £0\cdot8424$$

Variable cost/1000 components $= £842\cdot40$

These costs are plotted upon the break-even chart shown at Figure 14.24.

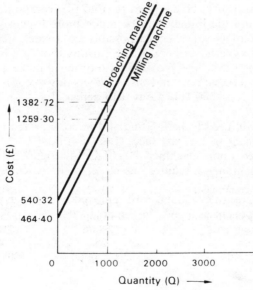

Fig 14.24 Break-even cost chart

In this instance both the fixed and variable costs of the broaching process are a little higher than for the milling process, therefore it is obvious that at the required production rate the milling process should be used. However, the break-even analysis is shown as an exercise. Had the components been required as fast as possible the situation would be quite different. Using more expensive tooling with a duplex fixture, the broaching process would be much faster than milling. Then a break-even analysis might well have favoured broaching due to the decrease in variable costs although fixed costs would have been higher.

(v) Having decided upon the milling process, the design features of the fixture can now be studied. These are (a) location, (b) clamping, (c) tool setting, (d) machine capacity and operator safety, (e) outline of construction of fixture.

(a) *Location.* The component can first be located on a pin of $15\cdot854^{-0\cdot05}$ mm diameter which allows any component bore within tolerance to locate easily and also allows adequate control over the $3\cdot18^{-0\cdot13}$ mm dimension. This location removes four degrees of freedom.

The fifth degree of freedom can be removed using two pegs as shown in Figure 14.25, which locate upon unmachined cast faces.

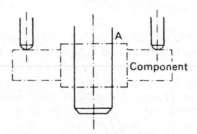

Fig 14.25 Locating component

These pegs will need to be floating and interconnected to allow for any mal-alignment of the cast faces. Location (and hence clamping) is not envisaged upon machined boss face *A* (Figure 14.25), as the support and clamping is required as close to the machining zone as possible for rigidity.

The sixth degree of freedom can be removed by location pegs underneath the component as shown in Figure 14.26.

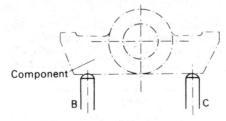

Fig 14.26 Locating component

These locators would contact the casting approximately on the mould parting line, and clearance may have to be provided on the pegs to give a solid seating. Any variation in casting size and shape must be accommodated. This could be done by making peg *C* fixed and peg *B* adjustable, this being designed so that the component is always pressed against peg *C*, either by automatic means or by the operator. The casting will then always be approximately symmetrical about horizontal centre-line.

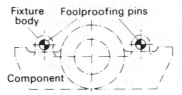

Fig 14.27 Foolproofing pins

All location surfaces should be hardened. Burrs from the milling operation must not prevent withdrawal of the component from the fixture. The component should be easy to unload provided peg *B* is easily retractable. Foolproofing pins will be needed, as shown in Figure 14.27, to prevent the component being loaded on to the main location pin upside down.

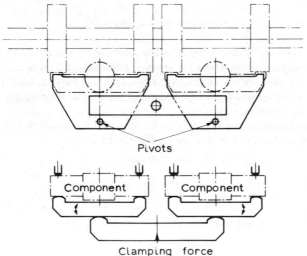

Fig 14.28 Clamping components

These pins can be accommodated in the main structure of the fixture.

(*b*) Clamping. The action of the clamps should be such as to thrust the component back against the location pegs which, being floating,

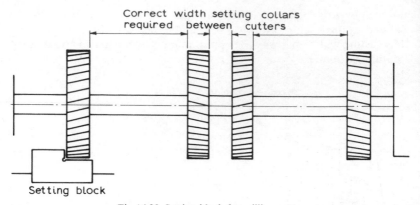

Fig 14.29 Setting block for milling cutter

would adjust themselves to suit any discrepancy in the casting shape. This arrangement should give maximum rigidity. Any clamps used should clear the cutters.

As two components are being accommodated in the one fixture, it would be desirable if the clamps could be actuated from a single control. Furthermore the clamps should preferably be floating to accommodate any discrepancy in casting size because they will be bearing upon a rough cast face. This can be done in the way shown in Figure 14.28.

The clamping force could be imparted by means of a screw or cam actuated by the operator, or could be pneumatically controlled. The final designed arrangement will depend upon the cost budget which was used for the costing analysis. At this stage only design principles are being committed to paper.

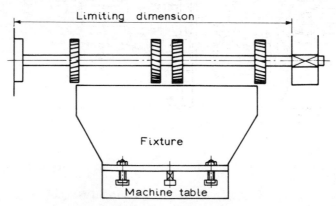

Fig 14.30 Fixture position on machine

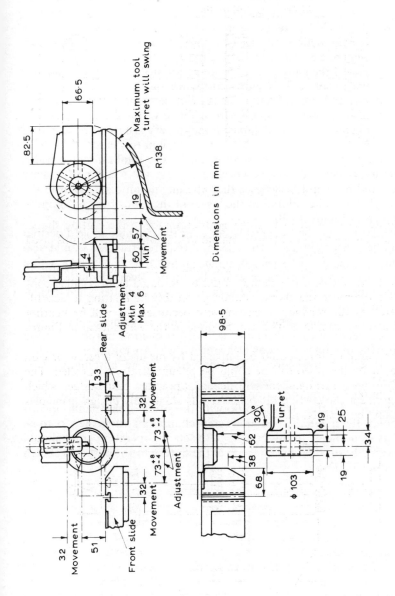

Fig 14.31 Capacity chart for automatic screw machine

In all clamping designs one should be careful that previously machined faces are not being marked. This is not a consideration here.

(*c*) *Tool setting.* A hardened and ground setting block is required, to be used in conjunction with feeler gauges (as shown Figure 14.11). In this case only one cutter needs setting in position relative to the fixture, then the rest of the gang of cutters are positioned by setting collars on the machine arbor. This is shown in Figure 14.29.

Hardened and ground tenon blocks will be required on the base of the fixture (as shown in Figure 14.12). These will make for quick and easy setting on the machine table although very accurate alignment of the machine and fixture axes is not necessary in this case.

(*d*) *Machine capacity and operator safety.* The total estimated width of the fixture should now be checked against the Machine Capacity Chart, and must not be greater than the maximum distance from the face of the machine column to the inside of the arbor bearing support, as shown in Figure 14.30.

The fixture may be designed wider than the machine table if desired using the construction shown. The total required traverse should also be checked against the stroke of the machine table. In this case with a traverse of only 35 mm, there would be a more than adequate stroke. Also tenon slot dimensions on the table should be noted when designing fixture bolt slots.

Accurate machine capacity charts showing every necessary dimension and detail of each machine used for production purposes should be filed in the Jig and Tool Drawing Office. An example of such a chart is shown in Figure 14.31, this being for an automatic screw machine.

When considering operator safety a difficult problem usually arises with milling fixtures. This is to do with the siting of the clamp operating point. If a manually operated screw clamping arrangement is to be used as seems likely in this case, then the clamp operating point should preferably be on the opposite side of the fixture to the cutter as shown in Fixture 14.32.

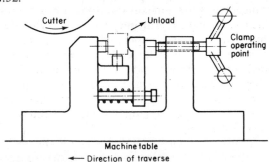

Fig 14.32 Clamp operating point relative to cutter

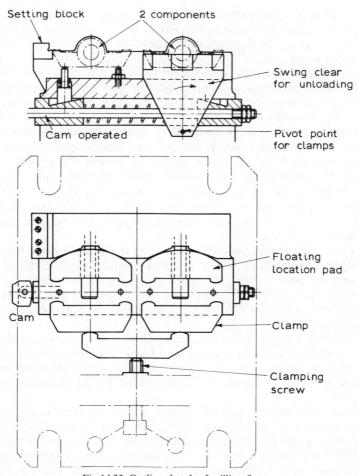

Fig 14.33 Outline sketch of milling fixture

The arrangement shown means that the clamps are taking the cutter thrust which is wrong in principle. However, with rigid and robust clamps this is not of vital importance. Remote control devices for operating clamps can be arranged if absolutely necessary; these are manipulated on the opposite side of the fixture to the cutter and the clamps. They will add considerably to the cost as they are quite elaborate in construction. Cutters must be guarded, and preferably a limit switch should be arranged to stop the cutters rotating when the table traverse is completed.

(*e*) *Outline of construction of fixture*. The main body of the fixture will best be manufactured from cast iron in the form of a casting for the reasons outlined earlier.

All the design ideas described up to this stage can now be put together and a picture of the milling fixture begins to emerge. This composite picture can be sketched to complete the design study and might look like Figure 14.33.

It will be noted that the location pins which removed the fifth degree of freedom have been incorporated into one unit which is floating and allows for variation in the size of the casting. There is now sufficient information in this design study to enable a jig and tool designer to complete the design on the drawing board and issue working drawings to the Toolroom where the fixture will be manufactured.

Standard forms could conveniently be used for design studies, a suggested one being shown in Figure 14.34, calculations and sketches accompanying the form.

	DESIGN STUDY SHEET	**Form No.**
COMPONENT	DRAWING No.	SKETCHES
MATERIAL	PRODUCTION QUANTITIES	
OPERATION	MACHINE	
LOCATION		
CLAMPING		
LOADING METHOD		
SECURING		
CONSTRUCTION		
SAFETY		
CUTTING TOOLS		

Fig 14.34 Design study form

Learning Objectives 14
The learning objectives of this section of the TEC unit are:

14. Evaluates the need for an application of jigs and fixtures to manufacturing processes.

14.1 Identifies and justifies the need for jigs and fixtures.
14.2 Identifies the difference between jigs and fixtures.
14.3 Explains the basic principles of jig and fixture design, e.g., location, clamping, construction and general design features.
14.4 Analyses the design of a given jig or fixture.

Exercise 14

1   How can the use of jigs and fixtures be justified?
2   With the aid of examples, explain the differences between jigs and fixtures.
3   What are the principal design points to observe for the location of a component in a fixture. Give an example of each.
4   What are the principal design points to observe for the clamping of a component in a fixture. Give an example of each.
5   A component 20 mm thick × 100 mm diameter is to have six equispaced 5 mm diameter holes drilled on an 80 mm *PCD*. Sketch the outline of a design of an indexing drilling fixture suitable for this drilling operation to be carried out on a single spindle drilling machine.
6   Sketch and describe two standard clamping devices in common use on jigs and fixtures.
7   Sketch and describe two location systems in common use on jigs and fixtures.
8   Using a component example of your own choice, carry out a simple design analysis for a drill jig, or a milling fixture.

Index